Also by Dr. Donald C. Johanson

Lucy: The Beginnings of Humankind
(with Maitland A. Edey)
Lucy's Child: The Discovery of a Human Ancestor
(with James Shreeve)
Blueprints: Solving the Mystery of Evolution
(with Maitland A. Edey)

JOURNEY FROM THE DAWN

LIFE WITH THE WORLD'S FIRST FAMILY

JOURNEY FROM THE DAWN

LIFE WITH THE WORLD'S FIRST FAMILY

DR. DONALD C. JOHANSON • KEVIN O'FARRELL

VILLARD BOOKS
NEW YORK 1990

Layout design: Kevin O'Farrell

 Published in the United States by Villard Books, a division of
Random House, Inc., New York, and simultaneously in Canada
by Random House of Canada Limited, Toronto.

Library of Congress Cataloging-in-Publication Data
Johanson, Donald C.
Journey from the dawn: life with the world's first family / by Donald C.
Johanson and Kevin O'Farrell.
p. cm.
ISBN 0-394-58084-2
1. Man—Origin. 2. Australopithecus afarensis.
I. O'Farrell, Kevin. II. Title.
GN281.J63 1990
573.3—dc20 90-50224

Manufactured in the United States of America
9 8 7 6 5 4 3 2
First Edition

To Lucy and the First Family, who were kind enough to become fossilized.

—DONALD C. JOHANSON

To my immediate family, Lori and Kim, and to the overall Family of Mankind on its journey through life.

—KEVIN O'FARRELL

When we stand on the Serengeti Plain, we are among trees, grasses, birds, and animals that evolved with us over millions of years—we are all *part of our natural world. With this perspective, it is so tragic that we, as a species—through technology and our never-ending appetite for expansion—can cause the extinction of even one of our fellow travelers.*

—DR. DONALD C. JOHANSON

PREFACE

Human beings have always been captivated by the beginnings of everything from science to art, but what fascinates us most is our quest to understand where we came from and how we got here. In pursuing our family origins, we are fortunate if we can look back a few generations. Yet the discovery of hominid fossils from Africa reveals that the roots of the human race can be traced back 4 million years. Based on these findings, *Journey from the Dawn* presents a tantalizing glimpse into the ancient world of our ancestors.

Paleoanthropology, the study of human origins, is a multidisciplinary science that attempts to discover the fossil evidence of our past. We use this evidence to reconstruct the anatomy and behavior of our ancestors and to understand the world in which they lived. Geologists study rock formations and reconstruct landscapes. They analyze volcanic rocks to determine the age of the fossils. Paleontologists study animal fossils and use them as clues to past climates and environment. Archeologists study stone tools, helping us to understand ancient behavior. Paleoanthropologists like myself study hominid fossils in order to understand their biology and reconstruct their physical appearance.

The striking resemblance between humans and African apes led scientists to speculate that they had a common ancestor and that the oldest evidence for our ancestry would be found in Africa. In fact, recent laboratory studies have stunned the world by revealing that chimpanzees and humans share 99 percent identity in their genetic makeup. These similarities can only be explained by the fact that sometime in the distant past humans and African apes shared a common ancestor.

It comes as no surprise that the most ancient hominid ancestor was found in 4-million-year-old African sediments. This hominid, known as *Australopithecus afarensis,* or more popularly as Lucy, gave rise to all later hominids. Some descendants, like the specialized vegetarian robust forms of *Australopithecus,* eventually became extinct. Another branch of our family tree led to a larger-brained form dubbed *Homo habilis* (the handy man), who began to make and use stone artifacts at least 2 million years ago. *Homo habilis* gave rise to *Homo erectus* about 1.5 million years ago and was the first hominid to leave Africa. Endowed with an even larger brain and possessing a body like ours, he not only manufactured stone hand axes but was the first to control fire. This set the stage for the emergence of *Homo sapiens,* our species, some 200,000 years ago.

This is just a thumbnail sketch of human origins during the last 4 million years, and many of the details continue to be debated. However, the evidence is irrefutable that the Human Family has a very ancient past. What is most intriguing about this picture is that the first appearance of *Australopithecus, Homo habilis, Homo erectus,* and even *Homo sapiens* was in Africa. There is no question that Africa is the cradle of humankind.

For the last twenty years, I have searched Africa's Great Rift Valley seeking fossilized clues to our past. The hardships of living in a tent in a desert climate where temperatures regularly soar to 110 degrees are far outweighed by the reward of finding hominid fossils that have rested in suspended animation for millions of years. Undoubtedly the most exciting moment in all of my years of exploration was the 1974 discovery in Ethiopia of Lucy.

Hadar, where Lucy was found, is a site located in a desolate region of Ethiopia known as the Afar Depression. Named after a local stream, the site resembles the Badlands of North

America. Centuries of water erosion have dissected these 3-million-year-old geological deposits. These strata are extremely rich in fossils of all kinds. In 1974, I was fortunate to be a coleader of an international expedition to explore this region of the Great Rift Valley.

Toward noon on November 30, a student and I were completing our morning fieldwork in an area at Hadar known as Afar Locality 288. While heading back to our Land Rover, I spotted a portion of arm bone lying on the ground. The anatomy of the bone was unmistakably hominid. We knelt down to take a closer look and noticed other portions of a skeleton nearby—a piece of lower jaw, a fragment of skull, the blade of a pelvis, part of a knee, ribs, and so on.

After our initial excitement subsided, we carefully located and collected all of the fragments of bone and teeth on the surface, recording their position on a map. Then all of the loose soil was sifted through fine screens so as not to miss even the smallest fragments. Following weeks of painstaking excavation we recovered hundreds of fragments, allowing us to piece together approximately forty percent of an individual skeleton.

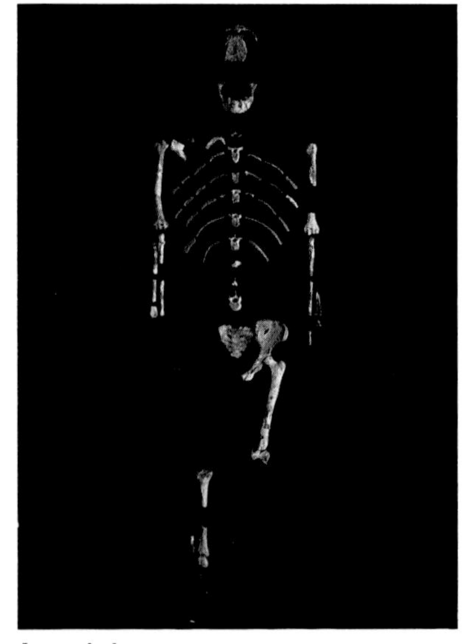

Lucy skeleton

Discovered in 1974 at Hadar, Ethiopia, this partial skeleton, known as Lucy, is the most complete example of Australopithecus afarensis *ever found. The 3-million-year-old skeleton is now at the National Museum in Addis Ababa.*

One evening, while celebrating this discovery of what would become known as the oldest and most complete hominid skeleton ever found, we were inspired to give her a nickname. Listening to the Beatles' song "Lucy in the Sky with Diamonds," out of sheer exuberance we called the skeleton Lucy. She also has an Ethiopian name, Dinquinesh, which means "thou art wonderful."

As there was no duplication in body parts, we knew we had the remains of only one individual. The small size of the skeleton told us it was a female, and the short thigh bone suggested a stature of about three and one half feet. The state of bone growth and the presence of wisdom teeth suggested that she died in her mid-twenties. Her cause of death has never been determined, but there is one tooth-puncture mark in her pelvis. Perhaps she was caught unaware by a crocodile while she foraged for food by the shore of the lake that had once been there.

During our 1975 explorations at Hadar, the field team located fragments of thirteen hominid individuals from a single fossil locality. We knew these fragmentary bones came from a single geological bed. There are remains of male and female adults and individuals as young as two years of age, known collectively as the First Family. How the catastrophic death of this 3-million-year-old group of hominids occurred is still not certain, but perhaps they were trapped by a flash flood.

Meanwhile, Dr. Mary Leakey's team at Laetoli in northern Tanzania discovered 3.5-million-year-old hominids that were very similiar to those found at Hadar. Her excavations at the site revealed a volcanic ash layer that preserved the footprints of birds, rhinos, baboons, three-toed horses, elephants, giraffes, dinotheres, cats, and hares, as well as a dung-beetle trail. Birds' eggs, bee and wasp cocoons, acacia twigs and leaves, animal dung, and termite mounds were also preserved. But most astonishing of all was a trail of footprints left by early hominids as they walked across the landscape 3.5 million years ago.

Taken together, the Hadar and Laetoli finds stimulated us to view human origins from a new perspective. After a long and detailed scientific study of these hominid fossils, a team of scientists came to a series of important conclusions: The hominid fossils from Hadar and Laetoli were indeed identical. Named *Australopithecus afarensis* in 1978, they represent the most primitive hominid ever identified. The pelvis, leg, ankle, foot bones, and especially the Laetoli footprint trail confirm that these creatures walked like modern humans. With a skull resembling that of an African ape and possessing a brain only a quarter the size of our own, *Australopithecus afarensis* looks like the "ape that stood up."

Recent discoveries indicate that *A. afarensis* roamed the African continent as far back as 4 million years ago. This highly adaptable species was at home in a wide variety of environments ranging from dry grasslands to humid forests.

Although the lack of artifacts at Hadar and Laetoli suggests that these hominids predated the invention of stone tools, it is possible that they made perishable tools of wood, grass, or twigs. Since our closest relatives, the chimpanzees, make and use rudimentary tools like termiting sticks, sponges to collect water, and rocks to break open nuts, we cannot rule out the possibility that Lucy may have used similar items as tools.

Clues to the behavior of Lucy's family are hard to extract from the fossil record. Studies of living primates do provide some insights: We know that chimpanzees, baboons, and apes live in large groups, so we can be fairly certain that Lucy's family was part of a much larger clan of hominids. Their social behavior was very different from that of modern human societies but was undoubtedly more complex than that of modern African apes. How widely they foraged for food each day, how large their territory was, what the details of their social structure were, as well as many other aspects of their behavior, will always be open to conjecture.

We will never be able to go back in time to observe our ancestors. We will always be detectives following the fragmentary trail they left. In reconstructing Lucy's family, one must not fall into the trap of considering them to be half human and half ape. *Australopithecus afarensis* was unlike any animal alive today, a creature as well adapted to its environment as any animal could be. Perhaps one key to its long existence was a high degree of behavioral flexibility, enabling it to survive in diverse habitats with a variety of strategies.

You may think we have put too much hair on their bodies or endowed them with too much cleverness or emotion. What you must understand is that this story is based on imagination, backed by scientifically grounded guesswork. We don't know that these specific events *did* happen, but scientific evidence tells us that they *could* have happened. We know that Lucy's family saw the cats, elephants, antelope, vultures, snakes, baboons, plants, and insects we depict—we have fossils, cocoons, and pollen as proof of their existence. Any geographical features and natural phenomena—such as lakes, rivers, rainstorms, and volcanic eruptions—were as much a part of their world as ours.

We will never know what thoughts trickled through those small brains 3 million years ago or what emotions the creatures felt toward each other. This is what makes exploring our origins so intriguing. By trying to define our place in the natural world, we are reminded of our humble beginnings, when we were intimately tied to that world.

In spite of our technological advancements, we are still very connected to the environments of our origins. Lucy reminds me of the delicate balance in nature. It is my hope that this perspective will encourage us all to guard that natural balance and in so doing assure a place for all life on this planet.

—Dr. Donald C. Johanson
April 1990

ACKNOWLEDGMENTS

The valuable insights into our earliest origins presented in this book would not have been possible without the generosity of the People's Democratic Republic of Ethiopia. We would like to express our sincerest appreciation to the Ministry of Culture and Sports Affairs for their encouragement and assistance. The fossil hominids discovered at Hadar are an important part of Ethiopian heritage. By making them available to the entire world, the Ethiopian government has made a contribution to all humanity.

We would like to express special admiration to Diane Reverand, whose enthusiasm, dedication, and brilliance have enriched this project at every step along the way. Diane went far beyond what is normally expected from an editor, for which we owe her our deepest gratitude.

Professor F. Clark Howell, of the University of California, Berkeley, was highly encouraging in the early stages of this project and we thank him for all of his scholarly advice. For valuable guidance on anatomical aspects of the early hominids, we are grateful to Dr. William Kimbel at the Institute of Human Origins. We wish to thank Dr. Robert Walter, also of the Institute of Human Origins, for his valuable geological advice. Thanks to Larissa Smith, my assistant, for her excellent editorial skills.

The photographs in this book were assembled from numerous sources. Special thanks to Carol Beckwith, Giancarlo Ligabue, Alex Marshack, Carlo Peretto, David Brill, and Hugo van Lawick for permission to use their remarkable photos. We appreciate Barbara Shattuck's help at *National Geographic* in obtaining the works of a variety of photographers. Thanks to the very helpful staffs of Peter Arnold, Inc., the Iwago Photographic Office, Bruce Coleman, Inc., Animals Animals, Photo Researchers, Inc., Robert Harding Picture Library, and Anthro Photo File for assisting us in finding just the right pictures. For discussions about insects and his wonderful photos we thank Edward Ross of the California Academy of Sciences. Thanks to Elizabeth Morales-Denney for her artwork.

I want to express special gratitude to my wife Lenora for her unfaltering encouragement throughout this project. Working together with my colleagues, Kevin O'Farrell, Harry Clark, and Reid Boates, has been a delight. I thank them for their stimulation and talent.

—DR. DONALD C. JOHANSON

I would like to acknowledge Diane Reverand, my editor but, in many ways, now a relative in my extended family. She saw this book as a positive force and has been continually supportive throughout its entire birth process. I would also like to thank my partners, Don Johanson, Harry Clark, Reid Boates, and Gerard Van der Leun, who all contributed their expertise. Emily Bestler and the rest of the Villard and Random House editorial, design, production, and marketing staffs have been more than professional. Bill Snyder, Brook Fancher, and Kathy O'Brien assisted in the background paintings of a number of spreads and my daughter, Kimberly, and wife, Lori, transferred many of my sketches to the painting surface. The photographers, too many to list here, provided rare glimpses of fascinations framed by their eyes. Elizabeth Morales-Denney provided the maps and scientific illustrations. I would also especially like to thank my inspirational parents, who followed a path of discovery from the remote farms of Ireland to the streets of New York and California.

—Kevin O'Farrell

INTRODUCTION

My experience in creating exhibitions for museums and other educational centers has led me to a realization that good education is more than knowing a lot about something. It is a game of applying that specific knowledge to the broader context of life.

Great science and great education should be considered as a convoluted series of interlinking mystery stories laden with clues. Each discovery merely points out the direction toward further clues and still further discoveries. Each step along the journey is a destination, if we are sensitive enough to perceive it as such. Each step is also an opportunity to discover something new or to select a new path, a new destiny.

I put my philosophy to the test when developing *Journey from the Dawn.* Lucy and the First Family are not presented as mere fossils but as living creatures strolling through their environment. It becomes readily apparent how interdependent they are with each other and with all the elements of their world. Each spread illustrates a threshold of discovery. The book's scientific text indicates some of the investigative paths that led to the story and others that may be derived from the story and applied in other spheres of learning.

My path in the creation of the narrative and illustrations for this book was one of continual discovery. Don Johanson, and scientists working with him at the Institute of Human Origins, helped shape the First Family of hominids from their amazingly complete fossil evidence. The bones defined not only the general appearance of the species, *Australopithecus afarensis,* but also the pronounced physical differences between adult males, females, and their children. Since no fossil impressions of soft body tissue and hair have yet been found, some educated guesses had to be made. These guesses were based upon the wealth of knowledge available on other primates and on other early hominids and primitive man. The painstaking reconstructions by anatomists Jay Matternes and John Gurche have been inspirational in these areas.

The personal names that I have bestowed upon the characters in this story just serve to distinguish one from another. They are not meant to convey any impression that this Family had a spoken language. All the names, except Lucy, have been derived from my own Gaelic tradition and reflect some characteristic of the individual's personality.

Don Johanson's and Bill Kimbel's studies of the brain capacity of the cranium of this species helped define their level of sophistication, which was not much greater than some other primates still in the trees. However, unlike the apes, these hominids were bipedal, which allowed their hands to carry food back to others in their family. These hands could also make and use tools. New opportunities challenged their evolving brains. Information about communication, family structure, and behavior was based on Don's research and on the studies of primates by Jane Goodall and Dian Fossey, as well as archeological evidence of early man and the work of anthropologists studying tribes of people living today in environments similar to those encountered by Lucy and the First Family. Books have been a major resource for my own research, but actual observation of the subtle behavior of primates and remote tribes from documentary television presentations has been invaluable.

Our journey of discovery led to Clark Howell's office at the University of California, Berkeley, for a survey of animals that the Family may have encountered. Some animals are extinct, others slowly evolved into different forms, while still others have remained unchanged in the past 3 million years. My visual reconstructions of some of the extinct creatures and descriptions of their characteristics are based upon fossil evidence compared to current animals that have adapted to similar life-styles.

A basic knowledge of the Family and the other inhabitants of their valley, together with its flora and geology, constituted my pallet for the story and illustrations. All those components were stirred together with time, seasonality, and environmental upheavals to form a soup of possibilities. We do not know if Lucy or her Family actually experienced any of the trials put before them in my story, but they most definitely could have. Our own future is a soup of possibilities based upon forces stirred from within and without.

—KEVIN O'FARRELL

JOURNEY FROM THE DAWN
LIFE WITH THE WORLD'S FIRST FAMILY

Life moved within her. Her eyes opened to a night sky shot through with a shimmer of stars and the arching plume of light that would someday be called the Milky Way. That day would not come for more than three million years.

For Lucy, such things did not matter. Her time was the unending rhythm of day and night, the migrations of the herds, and the beat of her heart. Lucy knew only of the warm night, the child growing within her, the bright moon making the savanna around the grove glimmer, and her daughter, Liban, whose small body nestled warmly against hers where they lay at the edge of the ravine.

The rest of her Family slept nearby within a grove of trees: her nine-year-old son, Lonnog, already grown taller than his mother; his father, Lorcan, the leader, who even now stirred under the trees; the elder female of the Family, Eba, old but still strong; and the others—Cano and Cera, Dagan, Brig, and Aine—together they roamed the vast valley that would someday be marked on maps as Africa's Great Rift Valley. All of them, Lucy's Family.

When the dark tentacle of a cloud snaked across the stars, Lucy shifted her head to follow it. The grass she lay on crackled, dry from months of drought. The cloud interested Lucy. It emerged from the top of the strange mountain on the horizon, thin and black in marked contrast to the building rain clouds in the distance. For as long as Lucy and the Family could remember, smoke and vile odors had come from the mountain.

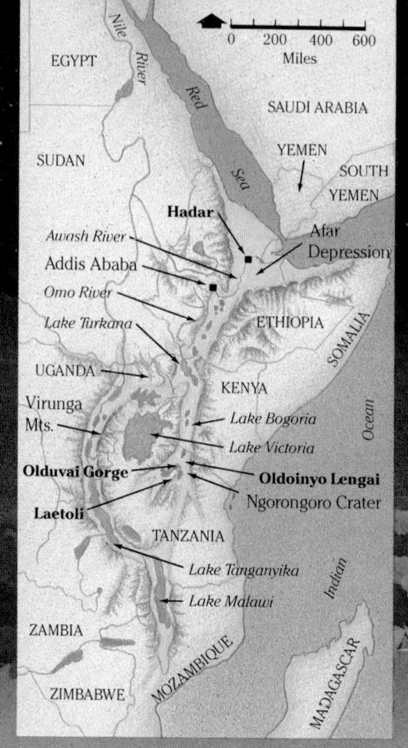

Africa's Rift Valley

The Great Rift Valley is a 4,000-mile fissure that extends from Mozambique into East Africa, up to the Red Sea, finally disappearing at Syria. The valley formed as a fault, or crack, in the Earth's crust, a place where lava from the molten mantle of the earth below squeezed up to the surface forming a chain of volcanoes. Constant eruptions of this magma, associated with the movement of the plates (huge sections of the crust), created extensive nutrient-rich plains. Seasonal monsoons from the Indian Ocean formed Africa's largest lakes and rivers. The resulting verdant valley of forests and extensive grasslands hosted the greatest concentration of land animals known, including early mankind and its ancestors.

The fossil evidence of Lucy and the First Family was found at Hadar in Ethiopia. The site lies in the Awash River valley which is part of the 58,000-square-mile Afar triangle area adjacent to the Red Sea. Some 3.2 million years ago, the area was dominated by a large lake with a delta of braided streams and islands on one side. The escarpment highlands were lined with ravines ending in alluvial fans, where sediment washed from the highlands was deposited. Marshes and forests surrounded the waterways and lake with plains or savannas extending beyond. Volcanoes and other geothermal activity dominated the landscape.

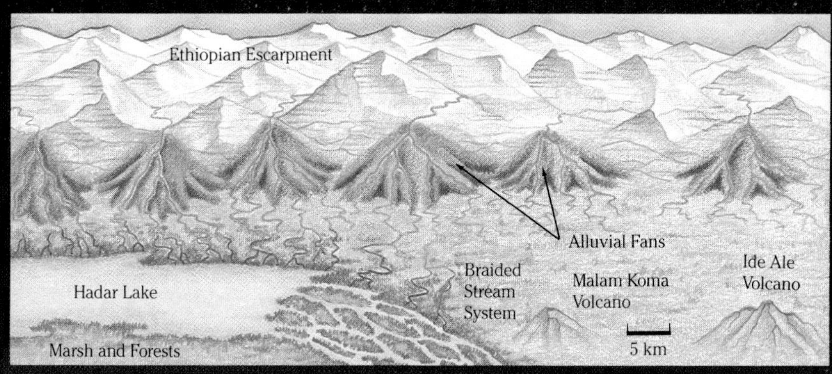

Hadar 3.2 million years ago

The Family felt the ground tremble on occasion, as if some giant serpent were struggling to shed its skin. Their fear had driven them away from the smoking black mountain. During the day, they had climbed from the valley to a hillside grove beside the vestiges of a stream that had drained the escarpment above it. Seasonal floods had sliced into the streambed, carving a deep ravine in the earth. Giant boulders, exposed by these forces of erosion, provided the Family with a fortress-like sanctuary with a ready supply of rocks to ward off attack.

The night was warm and close. Lorcan crouched on a ledge warily as the others slept below him. His deep-set eyes scanned the valley stretching beyond his lofty perch. His nostrils opened to the air, but there was no breeze to bring him scents. Sounds of the night echoed through the stone walls of the ravine. In the distance he could hear the splashing and periodic trumpeting of dinotheres, the roar of crocodiles in the marshlands, and the sounds of plains animals in a nightly dance of life and death. To Lorcan, the movements and sounds of the night were as natural as the rising of the moon and were not usually a threat to the Family. But on this night even the normal movements of the plain felt wrong. It seemed as if his entire universe, from the stars to the grass underfoot, stood poised for flight.

In their search for hominid fossils, paleoanthropologists travel to remote and desolate places in Africa's Great Rift Valley, where erosion has exposed ancient geological deposits. The tented field camp at Hadar, serving as home for fifteen scientists and students, was chosen because of its closeness to the Awash River, which provided a source of water for the expedition. The field team found Lucy in the surrounding fossil beds in 1974, and the First Family in 1975.

Vertical view of Hadar camp, 1975

The Dinotherium *was an elephant-like creature that ranged from Africa to Asia for 25 million years. It has been extinct since the late Pleistocene epoch, about 1 million years ago. Dinotheres are very similar to large African elephants in size and in outline except for peculiar down-turned tusks extending from the lower jaw.*

At the edge of the grove a large boulder jutted out of the grassy earth. Tightly grasping a thick branch he had carefully selected, Lorcan climbed silently to the top. A light breeze had come up, and sniffing the air, he caught a faint trace of smoke. He stretched himself up to his full height and grew as still as stone.

Lorcan looked over near the trees at the edge of the ravine where Lucy slept with their daughter, Liban. Still nervous, but seeing nothing to alarm him, he leapt silently from the rock and walked over to her. He knelt at her side. Lucy woke as quickly as a cat and looked into his eyes, making a soft sound in her throat that could not be heard more than three paces away. The little girl, Liban, woke too. Seeing her father, she

The 3.5-million-year-old knee joint (center), discovered in 1973, was the first hominid fossil found at Hadar. In the human knee (left) the shaft of the thigh bone rises at an angle to the horizontal, enabling the head of the thigh bone to fit into the hip socket. We walk with our knees close together so that they are directly under our body weight. The chimpanzee (a quadruped), walks on all fours with knees spread apart and the thigh bone at a right angle to the horizontal. The Hadar knee joint is virtually identical to that of a modern human, indicating bipedal walking.

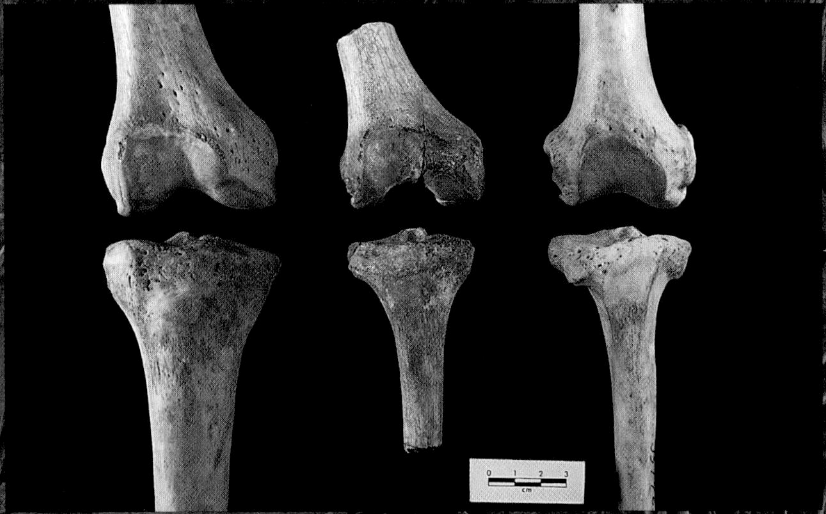

Comparison of human,* A. afarensis, *and chimp knees

withdrew a few yards, nestled in the grass, and went back to sleep.

Comforted by Lorcan's presence, Lucy reached up and drew his hand down to rest on her stomach. He stroked the soft fur there, feeling the strange, taut roundness of it. Suddenly, something within her pushed out. He jerked his hand back, startled. Then, contentedly, he gently touched her belly and resumed his stroking. The pushing came again. He looked at his mate resting calmly in the light from the moon. Sensing what the movement within her body meant, he felt deep within him that this night needed special vigilance. Kneeling there by Lucy he raised his eyes again, scanning the valley, watching for the slightest movement, for the barest hint of danger to the Family.

The plains below the hillside grove shone blue in the moonlight. The grazing herds, clustered across the plains, moved anxiously in and out of the shadows of gathering clouds. They seemed wary, but not just of the great cats and other predators that prowled the night. Instead, the herds seemed to draw in toward their centers, as if the threat came not from any one source but from all directions.

In the marshlands, still lush during this extended drought, the roars of huge pigs overpowered, for a moment, the continual rasping of the frogs along the bank. There was little wind to clear the dust from the air or to thin the lingering humidity of the day. Even the insects grew quiet as if responding to a silent warning.

Some distance away, a herd of wildebeest grazed nervously.

Wildebeest herd

Zebra and wildebeest

Tanzania's Serengeti National Park, a place of incomparable beauty and majesty encompassing 5,700 square miles, is one of the world's greatest natural treasures. This immense expanse of semiarid grassland and woodland is home to more than 1.5 million wildebeest, 200,000 zebra, a quarter million gazelle, and many other animals. The carnivorous lion, cheetah, wild dog, and hyena add drama to this stunning landscape. The annual migration of large herds of wildebeest and zebra following the rains in search of new grazing areas is awe inspiring. Each year thousands of tourists visit the Serengeti Plain and experience a world similar to that of our most ancient ancestors.

From time to time one or another would start at a shadow, or shy from a movement in the grass. A bellow would go up and the herd would shift for a moment before settling back to feed. Some would take a bite of grass and then raise their heads, ears alert for the slightest sign of danger. One of the older members of the herd was lagging slightly behind the rest, favoring a leg that had been scraped the day before. Behind it, a patch of tall grass moved as if blown by the wind—except there was no wind.

The giant cat vaulted out of the brush like a thrown spear. Days of failed hunting drove the megantereon at the solitary wildebeest with desperate speed. The wildebeest's head came up, instantly sensing the threat. It spun on its hooves, leaping away from the closing predator with a quickness that caused the cat to overshoot its intended kill.

Bellowing in panic as death closed in, the wildebeest bolted toward the startled herd, hoping to lose itself in the thousands of its kind and be passed over this night. But the herd was already stampeding away from the snarling cat. Cries and bleats rang out across the savanna as thousands of hooves thundered across the ground.

The wildebeest made a frantic sprint for the retreating herd. Then it felt the cat's teeth clamp down to the bone of its tail. The wildebeest's back legs buckled. It swung around, kicking with its hooves, trying to gore the cat with its horns.

The wounded wildebeest had almost succeeded in getting

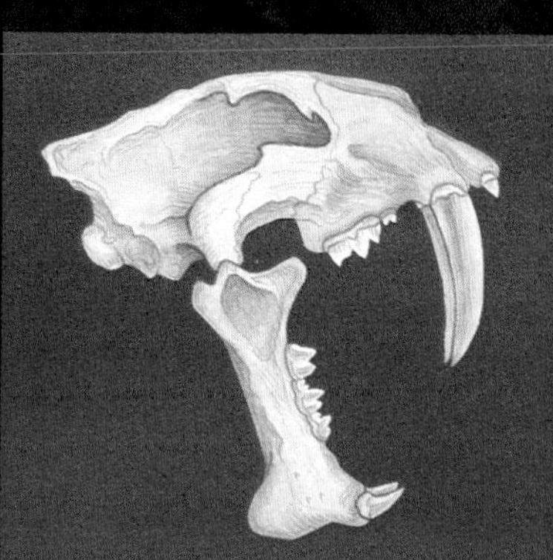

Megantereon *skull*

The canines of the Megantereon, *or dirk-tooth cat, were probably most useful in attacking the throat. The back teeth were thin and sharp and functioned like a pair of scissors for slicing meat.*

its horns into the attacker when a second cat surged up out of the brush and scrambled onto its back. The second cat's claws jabbed deep into the wildebeest's hide, dragging the wildebeest down onto the dusty earth. The first cat released the tail and, hissing, clawed over the struggling body to the exposed neck. Thrusting its head under the wildebeest's neck it pushed up until it felt the soft throat. Then its mouth gaped wide, the killing fangs gleaming in the moonlight, and drove deep into the wildebeest's throat in a spray of hot blood.

The wildebeest's eyes rolled up at the sky. The wound on its throat was torn wider and its body slumped into death. The dark blood streamed out onto the dirt and crushed grass. The cats snarled and hissed at each other. Then they slashed open the thick hide on the belly with their curving fangs, buried their muzzles deep within, and began to feed.

The hissing of the cats, the bellowing, the rumbling of the herd's hooves, and the scent of blood announced the fresh kill. Predators large and small began to move in, first the cats' cubs, then the males of the pride. Packs of lurking hyenas emerged from the night and skulked at the fringes of the kill.

The two females who had brought down the wildebeest tore at the carcass, stripping away the skin and chewing deep into the meat. Their cubs, dodging the cuffs from their elders, also claimed their gobbets of food, while the hyena pack circled the area looking for a way to steal their night's meal.

Rumbling snarls came from the grass as the male cats, content to lie back and let their mates make the kill, stalked forward to claim the carcass as their own. A sweeping paw knocked one of the cubs away as a roar made the rest of the pride back off. The males, snarling at each other now, hunched down at opposite ends of the kill to feed. The blood-drenched grass looked black in the moonlight as the males swallowed large chunks of warm flesh.

Spotted hyena yawning

Although we think of hyenas as cowardly scavengers, studies have shown that they are effective hunters. Starting at dusk, they hunt cooperatively in packs of up to thirty, chasing their prey. With massive jaws and bone-crushing teeth, they leave few leftovers.

The carcass was more than half-consumed when a single hyena dashed out of the dark and tore off a chunk from a haunch. The male cats whirled at this intruder only to feel the slash of teeth at their hindquarters as another hyena attacked the cats from behind. Angered, the cats swung again and chased the second hyena into the brush. In an instant, a dozen members of the hyena pack who had been slinking forward on their bellies sprang forward and covered the carcass, yanking off strips of flesh and swallowing them with ravenous, snarling sounds. The two hyenas who had distracted the cats streaked in from the night to claim their share.

Frustrated, the large cats loped back to the corpse but froze as the pack of hyenas rose as one from the kill and faced them, their yipping howls, raised hackles, and bared yellow fangs announcing their readiness to fight. As large as the male megantereons were, they were no match for the pack. Snarling deep in their throats, their carving canines bared, they backed away and faded like shadows into the grass.

Deep beneath the plains, white-hot magma had been pushing against the earth's crust for decades. Hot streams probed up, expanding hairline cracks, melting through layers of solid rock. For weeks small rivulets of lava bubbled up inside the crater of the volcano. Then the thick lava dome that held back the molten stone shattered and collapsed—but only for an instant.

Hunting dogs stretching

Nomadic wild dogs roam the Serengeti. Equipped with long, slender legs and powerful muzzles, these highly social carnivores hunt in packs. They can quickly break into a run of thirty-five miles per hour, ultimately bringing down their prey and tearing it to pieces. If they have pups, the adults will return to burrows, where the young have been hiding, and regurgitate food for them.

Black-backed jackal

The black-backed jackal, pictured here, has long legs, a narrow muzzle, and large, pointed ears. Weighing only about fifteen pounds, these small carnivores mate for life and rear their young together. Usually they hunt as a family, preferring small animals—newborn gazelle, birds, lizards, and insects.

Exposed to the air, the superheated gas blasted hundreds of tons of glowing lava into the night. Boulders trailing streams of fire vaulted over the land in long, burning arcs, exploding into blazing shards on impact.

Even some twenty miles away, where the Family had camped, the land swelled and lurched as if it were a table pounded by an enraged giant's fist. Glowing fountains of lava spewed from the ragged lip of the volcano and surged down its slopes onto the plains. Brush, grass, birds, and animals exploded into flame moments before the red walls of lava obliterated them. Glowing balls of incandescent pumice, blasted high into the atmosphere, began to rain down from the sky, igniting fires wherever they fell.

The scarlet column of fire and ash grew across the sky like some huge, terrible tree, its trunk pulsing black and red and orange, with branches of flame shedding leaves of white-hot pumice. Arcs of lightning crackled above.

It consumed the stars. It blotted out the moon. Its voice was a rumble in the ground and a shrieking roar in the air. Like an immense beast it rose up out of the earth. Its breath and touch scorched the savanna and strewed death and fire everywhere. Out on the plains, near the foot of the volcano, the skin of the earth split open, and lava bled up out of the ground and flowed across the flatlands. Every living thing turned and fled as the hot ash began to cascade from the sky.

Oldoinyo Lengai (meaning "mountain of God" in Masai) is located in northern Tanzania. Still active, the volcano is seen here during a major eruption in 1966. Prior to volcanic eruptions, the earth trembles, often producing dramatic earthquakes. The ground can swell and produce bulges. The earthquakes cause shock waves to travel through the ground, alerting animals to danger.

Oldoinyo Lengai eruption, 1966

Volcanic lava flow in Zaïre, Nyiragongo volcano

Crazed by the explosions and the quaking land under their feet, animals all across the savanna began to seek sanctuary. But sanctuary was not to be found. Driven by the fires, the herds fled across the plains, dodging flames that sprang up wherever the ash or burning rocks plummeted to earth. Ostriches screeched and ran from their clutch of eggs. Small dik-diks were trampled by bellowing herds of buffalo. The moonlight was extinguished by the plume of smoke from the volcano. In its place a red glow shimmered from the fire reflecting off the lowering ceiling of ash.

The wildebeest streamed across the savanna. Hipparion horses galloped in front of them and came to a halt at the steep walls of a gorge surrounding a stream. They milled around, then were driven into the water by the herd behind them. Screams of agony rose from the tangled bodies trapped in the water as limbs, backs, and skulls were crushed by those charging from behind. The stronger and luckier ones swam on and struggled up through the mud on the other side. The water was stained red by the fires above and by the blood of the dying animals.

Downstream, the river emerged from the gorge and flowed into the marshlands of the delta. Here too snakes, insects, and mammals large and small chose water over fire. All were swept away. Lifeless bodies snagged on rocks or spars, or rolled downstream to rot on the banks or be buried in the mud or under the red-hot ash.

Scores of crocodiles, piled on top of each other to stay warm, plunged into the bloodred waters. An antelope, both its forelegs broken by a fall, screamed as the jaws of a croc closed over its haunches and, with a harsh wrench of its body, tore it in two and began to feed.

Even dead, the animals were not safe from attack. As the carcasses began to wash farther downstream, they disturbed a group of hippos driven to the center of their domain by the noise and the smell of burning. Bewildered by the exploding night and the invasion of their territory, the hippos charged about the river, crushing the bodies and snorting furiously.

During wildebeest migration, the animals become so closely packed together that many panic and drown while trying to cross streams and lakes, providing a glut of food for nearby carnivores and crocodiles.

If the drowned animals become covered by mud, sand, and silt, their bones will be gradually transformed into fossils. This may result in a compacted bone bed, like the one pictured here at Isernia la Pineta, a 700,000-year-old site in southern Italy.

Wildebeest drowning

Compacted fossils at Isernia la Pineta, about 700,000 BP

The jarring blast of the eruption snapped the Family awake. They impulsively began to run in all directions. Then they stopped and pulled back into a group. The grove was burning. Eba, the oldest female, began to urge the other females and the children toward the ravine behind the camp.

The adult males, Lorcan, Cano, and Dagan, lagged behind for a moment as if to defend their mates and children from this sudden danger. The air became hotter as they looked to all sides. A shaft of fire shot through the sky and exploded on the ground in front of them, setting the brush on fire. The three males scattered at first, but then Lorcan curiously approached the flames. His lips curled back in a fearful grin as he stepped cautiously toward a stick that was burning at one end. He budged it, then hooted and leapt away. Again he approached and grabbed at the stick. He held it up and looked at the burning tip, then at the pillar of fire that was shooting up from the volcano. He touched the flame with his finger and gave a yelp, dropping it on the ground and jumping back. Cano leapt forward with a large stick and clubbed the offending branch over and over again until all the fire was driven out of it. Lorcan stepped forward and pushed Cano back as he picked up the charred stick and, hooting, waved it defiantly at the volcano. At the same time, the fire that had started in the brush had reached a dry acacia tree and started to roar up its trunk. Seeing the fire spread, Lorcan turned and, followed closely by Dagan and Cano, scrambled over the edge of the ravine. In the darkness, they failed to notice the small form of L[illegible]y, who had stopped a moment before following the rest of the females.

Humans belong to a group of animals called primates. Primates consist of two major subdivisions: the prosimians and the anthropoids. Prosimians include lemurs, lorises, and tarsiers, which resemble some of the earliest primates that lived 50 million years ago. To a large extent, prosimians are nocturnal and eat insects and small reptiles.

The anthropoids, including monkeys, apes, and humans, are active during the day and have binocular, color vision.

Lucy trailed behind the others, confused and frightened, the movement within her growing more distinct and painful. Clutching her stick, she pulled herself up and moved quickly to the edge of the ravine. Awkwardly, she climbed down the sides to the bottom. Behind her, hot blades of fire stabbed up into roiling clouds of smoke. The dying screams of burning animals pierced the night. She hooted out into the ravine where the others had gone. There was no trace of her Family.

She sniffed the air for their scent. There! Using her stick for support, she staggered after them.

She had taken only a few steps when the pains from her womb racked through her again. She remembered Liber, the little male who had died the day after she had given birth to him a year before. She knew that her time was near again. She knew that she must get to her Family.

Lucy howled into the night. As her call faded, she heard a high-pitched hooting. It was her mother, Eba. Lucy howled again. This time the answer was closer. Then Eba's arms clasped Lucy, and she began to help her away from the fire.

Chain lightning snapped through the clouds. The flashes illuminated the ravine in brilliant frozen moments. The thunder shook their bones. Sheets of hot ash driven by a rising wind swept over them, stinging their bodies and driving them on.

Around a bend, Eba and Lucy stumbled through a waist-deep pool of cold water made murky by the falling ash. They clambered out of the pool and moved up a slope toward some rocks. Lucy sagged with effort and shook with the pain of the contractions. At last, in a nave in the rocks, Lucy dropped to her knees and gestured that she could not go on. Eba tugged frantically at her daughter's arm, urging her to rise. Lucy could not.

Instead, she keened a low sound, a sound Eba knew as the call for Lorcan. Lucy pushed at Eba, urging her to find Lorcan. Lightning flashed again as Eba, moving as fast as she could manage, clambered along the stream after the Family.

Lucy watched Eba as the night closed around her. Here, next to a lone tree, deep in the ravine, she would be safe for a time. The fires above burned fitfully and had not found their way into the stony ravine. They were still distant but moved toward her through the constant rain of ash that fell from the sky.

Lucy clutched the tree, its solidity reassuring her. Exhausted, she stared into the darkness. The racking pains came again, and Lucy waited alone in the night.

Most primates live in complex societies and have elaborate ways of communicating with each other. They employ not only vocalization, but also postures, gestures, and facial expressions. Loud cries are used by monkeys and apes to announce the presence of food. Certain vocalizations are alarm calls. Vocal and visual signals are used to stay in contact and to convey feelings and moods. Also an important means of communication, touch serves to convey reassurance.

With her extensive studies of chimpanzees, Jane Goodall has shown that they have a remarkably complex set of visual, vocal, and tactile signals. Although chimps lack language, they can communicate a wide range of emotions and moods. From such observations, we can surmise that Australopithecus afarensis *was capable of at least the level of communication seen in chimpanzees.*

The Family emerged from the ravine before the dawn. They scrambled to the top of a large boulder next to the stream and hunched down, sniffing the air. It was thick with the smells of burning and death. The explosions from the volcano had become less frequent. Still, rivers of lava crept across the ruined flatlands sending vast clouds of hissing steam high into the dusty air.

Shivering, the Family looked to Lorcan, but Lorcan only stared back up the ravine. He knew that Lucy was somewhere behind them.

Lorcan looked at the members of the Family gathered on the rock. Lorcan's brother, Dagan, huddled with his mate, Aine, and their younger children Aife, Ocan, and Ciar. Lorcan's daughter, Liban, sat quietly next to Cano and Cera, who were childless. Brig, Dagan's sister, hugged her two sons, Baccan and Buan, tightly. Lonnog and the teenagers, Bran and Anu, perched above the others on a great stone. Only Lucy and Eba were still missing, somewhere back up in the ravine.

Chittering among themselves, the Family waited. Far away, a muffled rumble came to them through the clouds. Another, close this time. Then a series of bright flashes slashed through the gloom.

Cano jumped back as a large drop of water splattered on his head. Then another came, and another. Soon a torrent fell through the ash and smoke, soaking their fur, washing away the caked ash from their skin.

The Family rose up as one when the advancing bolts of lightning stabbed across the sky over the devastated plains that lay before them. Those that carried sticks hooted, screeched, and brandished them at the sky. Cano shook his stick at one loud thunderclap. Others leapt from the boulder and began to prance about on the ground, their feet splashing in the silt and the mud. Their arms upraised and waving in the rain, they turned and spun, gesturing upward, acknowledging the forces above.

Jane Goodall has observed chimpanzees in what is described as a "rain dance" during heavy downpours. Males run, slap the ground, brandish branches, and bark loudly. The females and juveniles avoid the display by climbing into trees where they sit and observe the larger males. Amidst the thunder and lightning, the males hurl themselves up and down grassy slopes. After about an hour, the display stops and the chimps return to calmer, normal behavior.

Chimp in rain

Lorcan too leapt from the rock, but not to dance in the rain. Instead he moved purposefully back into the gorge. Cano and some of the young males turned and began to follow him. He faced them, growling and gesturing with his hands, and made them understand that they were to stay with the rest of the Family. Turning, he sprinted up the gorge, leaping from rock to rock until the Family could no longer see him. The rain became torrential, and Cano and the males continued their strange dance.

Lucy had squatted down beside the tree and howled as the pains came through her in waves. Then they passed and, panting, she opened her eyes to see Eba, her mother, with sweat and rain coursing down her face and matting the gray and brown hair.

Lucy closed her eyes again for a second and, as the next contraction made her body shudder, pressed back harder against the tree. The baby moved and shifted in the birth canal, struggling to be born. Lucy opened her eyes, stared blindly up at the rain above, and pushed. Her body shuddered with pain, but she made no sound. Eba's hands reached out and she made soft, encouraging sounds. Lucy pushed again with a hissing moan as the damp crown of her child's head emerged from her body. Another instant and the head was free, cradled gently in Eba's hands. Then the body of the child followed, sliding into the world and crying out as the rain washed it and Eba enfolded it gently into her arms.

Lucy slumped back against the tree, her breath ragged and panting as the pain of birth receded. Eba leaned against her, their two bodies forming a roof to protect the new child from the rain that began to sweep through the gorge and swell the stream below them.

During birth, both the mother and her offspring are highly vulnerable, especially on the grasslands of Africa where hungry carnivores are always vigilant. Wildebeest protect calves by giving birth in groups of twenty or more animals. The females not giving birth chase away predators.

Most wildebeest births occur within a very short three-week period. Because predators are quickly satisfied with such an excess of food, many calves escape.

Primates that are active during the day, including humans, prefer to give birth at night, avoiding potential predators. Primates apparently do not like to be observed giving birth. The same stages of birth seen in humans are typical of all primates. Nonhuman primates like monkeys and apes have a shorter period of labor, and this may also have been true for Australopithecus afarensis.

The pelvis is one of the most diagnostic portions of the skeletal anatomy for evaluating the mode of locomotion in A. afarensis. *The remarkable similarities between the carefully reconstructed pelvis of Lucy (upper left) with that of a modern human female (lower left) are obvious. The anatomy of these two bipeds is dramatically different from the high, narrow pelvis of the quadrupedal chimpanzee.*

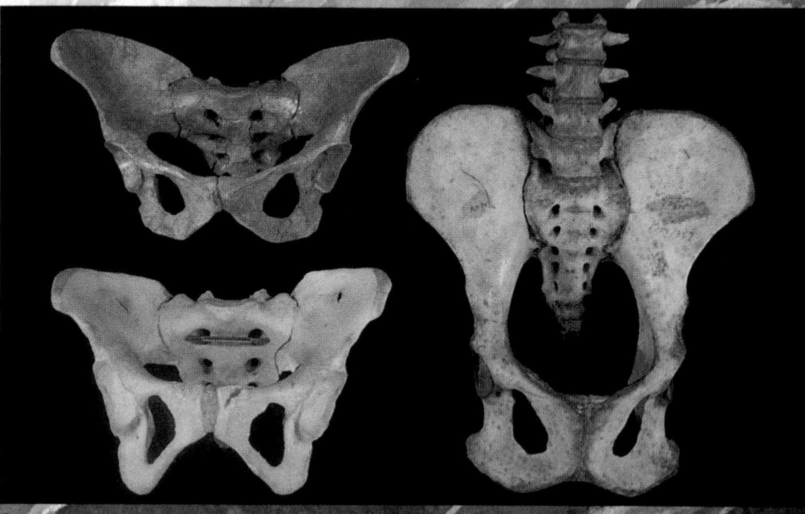

Comparison of Lucy's, human's, and chimp's pelves

The newborn infant squirmed against Eba as the storm intensified. A bolt of lightning directly overhead made the gloom of the rain fade for an instant and illuminated everything around the two females in sharp relief. A series of staccato howls from the gorge below made them peer through the sheets of rain.

It was Lorcan splashing toward them up the ravine. He gained the near bank and staggered up, water streaming from his body. Lorcan's lips drew back and he howled a warning. He pointed to the stream, a torrent filled with brush, logs, and rolling boulders. As they watched it flowed higher still, and then they heard a dull roar echoing down the ravine. He pointed above them. Understanding, Eba turned with the baby held tightly against her. Lucy turned as well, and they all began to scramble up the slope. The earth beneath their feet trembled as a living wall of water swept into the gorge.

The crest of the flood was jet black with ash. Whole trees rolled within it. The ripping sound of huge stones pushed along at its base echoed off the walls of the gorge. The raging flood almost reached them, but they struggled above it at the last moment, and it passed below, leaving behind a landscape scoured clean of all life. Only stones remained—stones and large pools of thick, black mud. The small ledge and the tree where Lucy's baby had come into the world had vanished.

The discovery of the First Family at Hadar in 1975 represents one of the most important finds ever made. Finding a hominid fossil is in itself a rare occurrence, but to have found more than two hundred fossil hominid bones, comprising at least thirteen individuals, is amazing. Close examination of the locality where the remains were found indicates that the accumulation was due to a catastrophic event. This is surmised because only hominid bones were found in this collection. If the bones had accumulated over a long period of time, other animal bones would have been found nearby. Carnivores were not responsible for the collection, because there are no characteristic chewing marks on the bones. Australopithecus afarensis *lived near water. It is possible that this group may have been drowned in a flash flood, or become bogged down in mud, caused by a river overflowing its banks. Further excavation may provide more clues as to why these hominids came to rest in a common fossil grave.*

Below them, tumultuous cascades of dark water coursed through the ravine. Still sheltering the infant, the three adults caught their breath as the rain began to lighten. The scent of smoldering ash gave way to the refreshing aroma of moist earth. Lorcan stepped away from the group and shook himself, a spray of water flying from his fur. Handing the child to Lucy, Eba did the same. Lucy, secure now at the top of the path, settled down on her haunches and gently began to lick her new baby clean.

The infant, born earlier than most babies of the Family, was tiny but perfectly formed. The baby was, as Lucy now saw, a female, and her eyes were closed in sleep. Her smooth face was, like most of the children, much lighter in tone and less defined than those of her parents. Lucy held the child out for Eba and Lorcan to see.

Long after, Lorcan and Lucy and Eba and all the members of the Family would die; long after, their bones would be dust or fragments of fossil washed by other rains; long after, the forest and plains on which they lived would turn to desert; long after all of this, what they felt gazing down at the sleeping infant would be recognized as the wonder at the promise of a new life. The child, Lifi, was that promise.

Holding the new baby close, Lucy and Eba followed Lorcan down the edge of the ravine toward the rocks where the Family had been left.

Cheetah family in rain

For most animals, the key to success is the availability of water. A hot, dry climate with seasonal rain typically produces areas of open vegetation like the Serengeti savanna, dominated by thorn trees called acacias and large expanses of grass. The two normal rainy seasons are in November through December and in March through May.

Vegetation on the Serengeti is well adapted to dormancy during the long, dry season. In a matter of days, a virtual desert is transformed to plains of lush, green grass—some 1,500 tons per square mile.

Lorcan, with Lucy carrying her newborn, clambered wearily down the rocks. Eba, who had twisted her ankle, limped behind. As they moved, Lucy walked as if asleep, her head nodding forward every few steps, then jerking upward.

As they rounded the last boulder, the area below them where Lorcan had left the rest of the Family was a scene of terrible devastation. The giant boulders had been splintered by the force of the flood. Shattered trees littered the sides of the stream. There was no sign of the Family.

Lorcan called out. His screech rose to a long howl and fell sharply silent. There was no response. He called again, louder than before. Nothing. Frantic, he climbed to the top of a mound of muddy rubble and shrieked again across the wasted landscape. Except for the rushing of the waters below, everything held its silence.

Eba moved painfully down to the water's edge. There a large tree had wedged itself between two stones to bridge the stream. Examining it and the stones beyond, she saw that it was possible to cross the stream, and hooted excitedly. Lorcan leapt down from his perch on the rock to stand beside her. Then Eba gestured to the slope on the far side of the stream. There, still visible in the wet earth, were the tracks of the Family leading toward the unburned top of the escarpment.

It was all they needed to see. Lorcan led them across the natural bridge and up the escarpment. There, at the edge of a grove in the distance, they saw the Family sitting wearily under the trees. This time Lorcan's call brought a host of answering calls. One small member broke away from the group and came running to the arms of her father.

Liban chattered noisily and wriggled out of her father's hold to dance around her mother and Eba. She stopped as she saw, nursing at Lucy's breast, a very small creature she did not recognize. Liban shook her head and spun away from her mother, not understanding where this new small being had come from. Her mother had not had it at the start of the terrible night, and its appearance now made the day almost more terrible than the night.

Lucy took Lifi from her breast and handed the protesting infant to Eba. Then she folded Liban into her arms and stroked the child until her head-shaking stopped. She took Liban's arm and put her hand on her flat stomach, then on the infant that Eba held. She pushed Liban gently forward, encouraging her to smell and touch the new infant. Slowly, Liban began to understand that this young one was also of the Family.

Some investigators believe that one critical factor in early hominid success was male-female pair bonding. Male involvement in the care of offspring, and recognition of paternity, led to a unique social unit: the family. The development of the nuclear family may be one of the key adaptations which contributed significantly to the success of hominids.

After birth, chimpanzees are highly dependent on adults, particularly their mother, for a relatively long period of time. A chimp isn't really grown up until it is ten to fifteen years of age. A baboon reaches maturity in only seven years. Born into a fairly complex society, chimps have a great deal of learning to do in order to function properly in that society.

A young chimpanzee will learn from its mother where to look for food and what sorts of things are edible. Chimps and other primates are not born with a set of instincts which assure their survival. In order to survive, chimps must learn what dangers there are and how to avoid them. In the larger framework of the troop, a young chimp will learn how to recognize the moods of other individuals and how to communicate its own emotions. Without childhood learning they would be incapable of living successfully in a troop. There is no question that the complexities of Australopithecus afarensis *society also demanded a prolonged period of childhood dependency.*

Chimp mother and offspring

The floodwaters, surging from the ravines into the river, had carved a broader and deeper channel. They had stopped the advance of the fires and had given the valley a strange and eerie look. On the one side, there was singed earth dappled with muddy ponds, gray slag, and black tongues of hardening lava. On the other, the gray mud of the river's banks yielded to charred grass, then to unburned groves and swaths of vegetation on which the surviving herds of animals grazed.

The rain quenched the fires in the brush and trees of the great plain. Sweeping from the headlands, floods scoured the land. Where the waters met the lava, the molten stone congealed on the surface, but its deep and abiding heat still threw up great clouds of white steam that dotted the landscape for miles around.

After reassuring Liban, Lucy took Lifi back and continued to nurse the infant. Finally, Lifi was content and, asleep, was handed into the arms of Eba to be presented to the Family. Eba was the oldest of all the Family, the last of her generation. It was rare for any member of the Family to see a grandchild. Life on the plains was usually much too short. Yet Eba

THE FIRST FAMILY

Danu *F-30* — Artan *M-32*
- Brig *F-38* *(Broc)*
  - Baccan *M-3*
  - Benen *M-9*
  - Bran *M-13*
  - Buan *M-5*
- Dagan *M-30* *(Aine)*
- Lorcan *M-35* *(Lucy)*

Eba *F-60* — Enan *M-64*
- Lucy *F-28* *(Lorcan)*
  - Liber *M-0*
  - Lonnog *M-9*
  - Liban *F-4*
  - Lifi *F-0*
- Aed *M-42*
- Barr *M-35*
- Cera *F-27* *(Cano)*
  - Corc *M-0*

Baine *F-35* — Cian *M-52*
- Cano *M-40* *(Cera)*
- Conn *M-12*
- Aine *F-28* *(Dagan)*
  - Aife *F-4*
  - Anu *F-14*
  - Fann *F-10*
  - Ocan *M-5*
  - Ciar *F-7*
- Broc *M-44* *(Brig)*

NOTE: Names in parentheses are mates.
Names in boxes are deceased. The number indicates their age at death.
Sex and age are indicated below names.

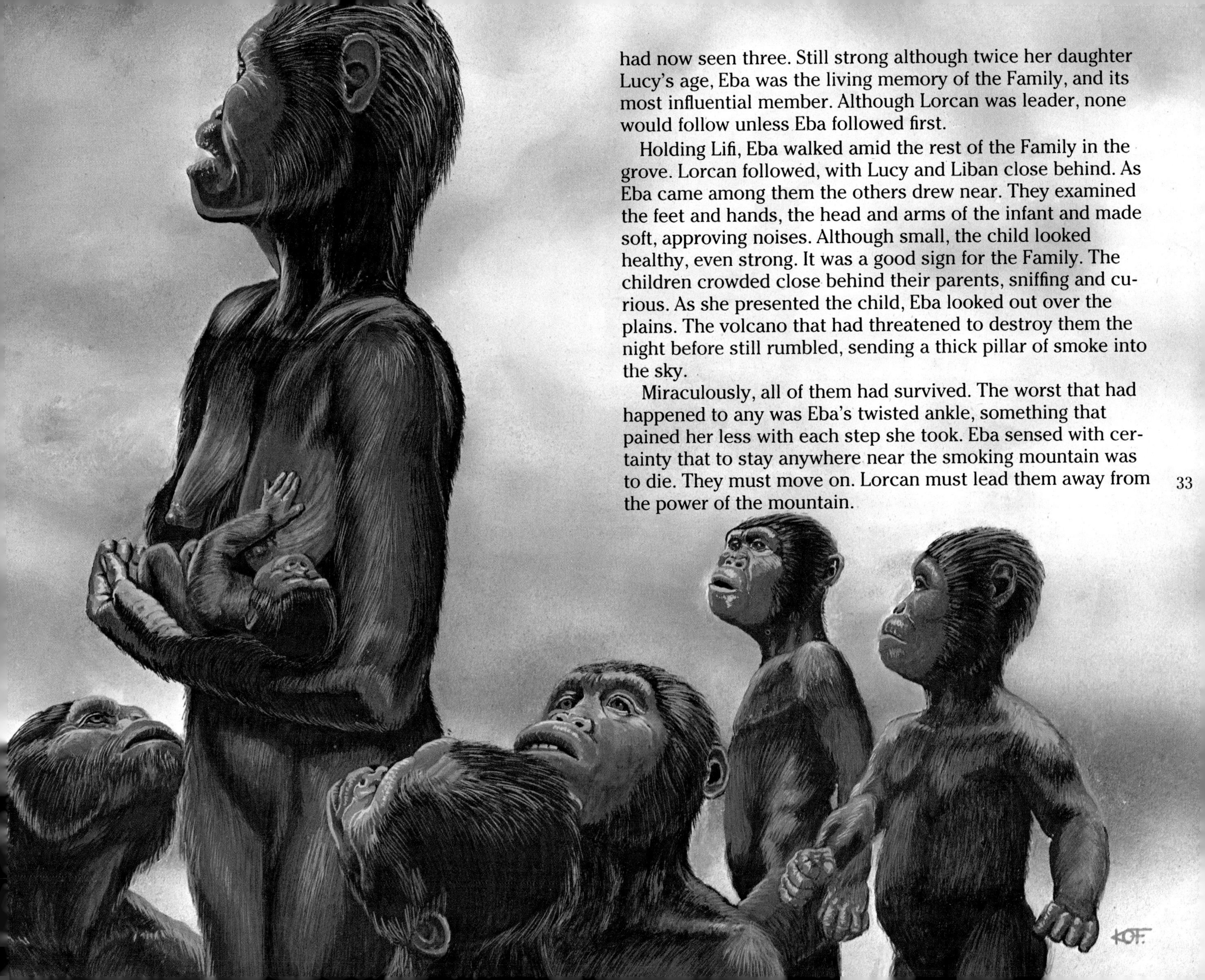

had now seen three. Still strong although twice her daughter Lucy's age, Eba was the living memory of the Family, and its most influential member. Although Lorcan was leader, none would follow unless Eba followed first.

Holding Lifi, Eba walked amid the rest of the Family in the grove. Lorcan followed, with Lucy and Liban close behind. As Eba came among them the others drew near. They examined the feet and hands, the head and arms of the infant and made soft, approving noises. Although small, the child looked healthy, even strong. It was a good sign for the Family. The children crowded close behind their parents, sniffing and curious. As she presented the child, Eba looked out over the plains. The volcano that had threatened to destroy them the night before still rumbled, sending a thick pillar of smoke into the sky.

Miraculously, all of them had survived. The worst that had happened to any was Eba's twisted ankle, something that pained her less with each step she took. Eba sensed with certainty that to stay anywhere near the smoking mountain was to die. They must move on. Lorcan must lead them away from the power of the mountain.

The Family needed to rest. The night had been exhausting. Lucy was still weak from childbirth. Cano was already in a deep slumber, while Dagan nodded over his sleeping mate, Aine. All nine children were napping. Lorcan too showed signs of deep fatigue, but he would watch awhile longer to make sure the burnt grove they were in was safe. Eba sat cradling the sleeping Lifi in her arms as the Family slept about her. The sun that had just poked through the clouds was warm on her body as she drifted into sleep.

The sound of quiet sobbing nearby and the touch of Lucy's hand on her arm woke the old female. It was Lucy's sister, Cera, her face full of tears as she sat near Eba and stared at Lifi. Cera had been the last female to give birth. Cera's baby had died in her arms a few minutes after being born. At first, Cera's wails at the death of her baby had shaken the air. Refusing to accept that her baby was dead, she had carried it with her for three days, trailing distractedly behind the others.

When her mate, Cano, had once attempted to take the infant from her, she struck him on the head and screeched. After that, Cano and the rest of the Family watched Cera carefully but did not interfere with her grief.

On the third day after the baby's death,the Family had

As might be expected, primate mothers grieve when their offspring die. Macaque monkeys have been observed carrying a dead infant around for days before abandoning the carcass. In instances where an infant loses its mother, the infant may die. Jane Goodall has shown that among chimps, sometimes an older sibling or an aunt will adopt the orphan and look after it. In macaque monkeys, infants who had lost their mothers were sometimes adopted by adult females. In fact, the infant itself usually chooses the new mother.

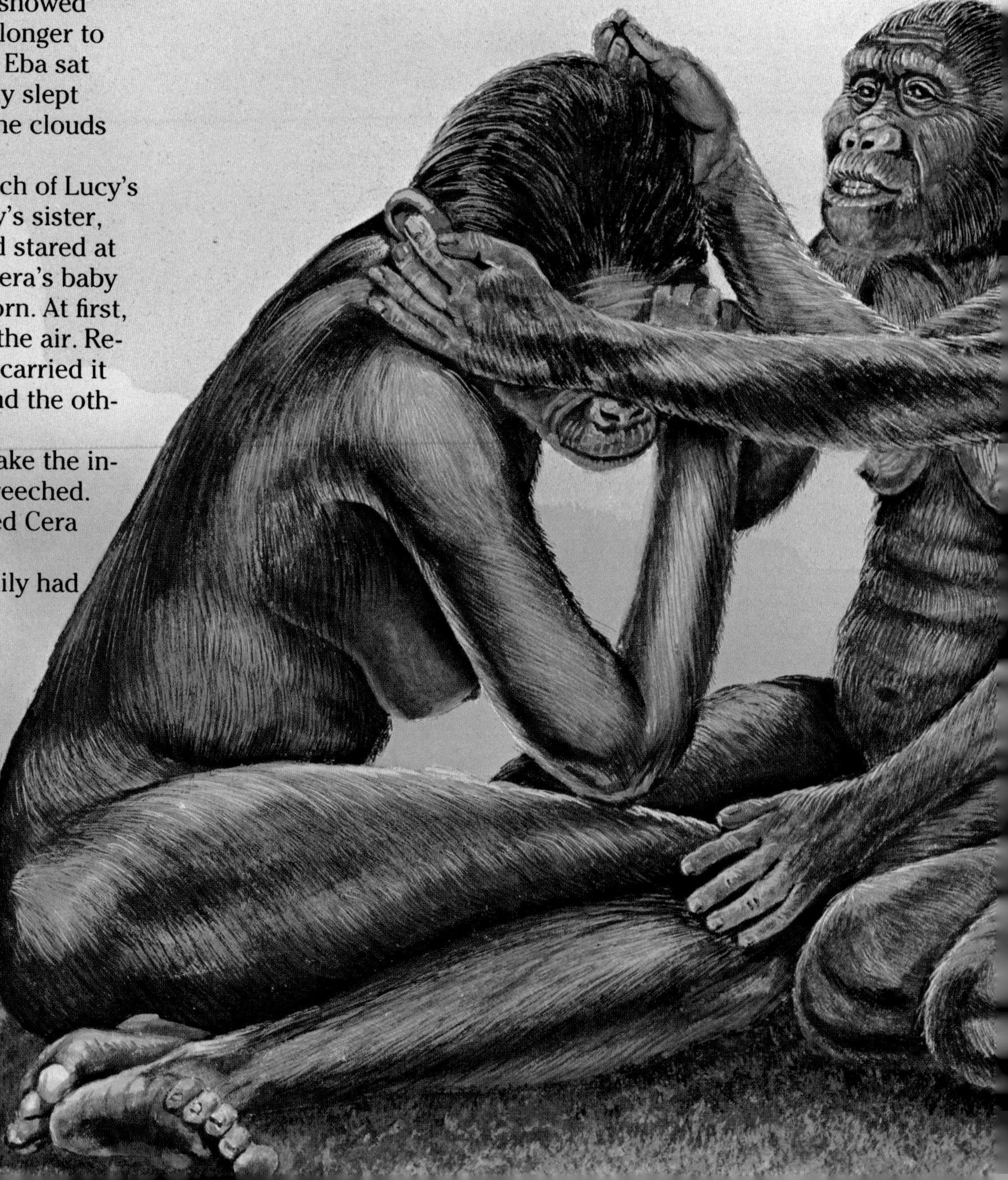

stopped by the edge of a large lake. In the last light of day, they watched as Cera made her way to the shore and laid the body of her child gently amid some large stones. She turned away and sat for the rest of the night apart from the Family. In the morning, they left that place.

In the weeks since, Cera had held herself apart from the Family. In Lifi, though, Cera found something of the infant she had lost. Eba handed the baby to her grieving daughter. Then she and Lucy sat near Cera and began to groom her gently. Liban, Aine, and Dagan's four-year-old daughter Aife stood by watching. After a time, Cera rose and went to sit by her mate, Cano, again ready to be part of the Family, ready to go on.

Chimpanzees exhibit a wide range of emotions. On the left, a chimp is showing fear; those on the right are showing affection by kissing. Depression can have a devastating effect on any animal, and chimps are no exception.

Jane Goodall, who has studied chimpanzees in their natural habitat for more than twenty-five years, reports how serious depression can be among these animals. One of her study chimps, Flint, had a very special relationship with his mother, Flo. When she tried to wean him at the customary age of three, Flint refused. Flint continued to stay close to his mother, sleeping in her nest until he was eight. When Flo contracted flu and died, Flint became very depressed and his own health began to fade. Flint sat for hours next to his dead mother and even his sister and brother couldn't cheer him up. Finally, after three-and-a-half weeks, Flint died, probably from diseases aggravated by depression.

Chimpanzee showing fear

Chimpanzees showing affection

Midmorning found the Family at the edge of a vast expanse of muddy ash. Beyond it, Lorcan could see a faint haze of green where the plain began to live again. He raised the stick he carried and gestured to Cano and Dagan, the youngest male of the Family. Cano moved back and to the left of the females while Dagan shifted to the right. Then Lorcan stepped forward into the mud and moved off toward the green tree-tops on the horizon.

His eyes scanned the vacant landscape for predators, his stick held low and at the ready. The other two adult males flanked him. After a few minutes in the ash wasteland, Lorcan began to relax his vigil, and the children of the party scampered in front of him. Lonnog ran ahead, followed by Brig's three sons by Broc, who had been killed the year before. Baccan was three, Buan was five, and Bran, at thirteen, was almost an adult. Behind Lorcan, Aife, Anu, Ocan, and Ciar stayed close to the females. Following behind was Lucy, with Lifi held close, and Cera nearby. Then came Brig, Dagan's mate, supporting Eba, who limped heavily, her ankle again hurting her.

Lorcan felt a tug on his arm and looked down to see Liban gazing up at him. She raised both her arms. Lorcan smiled and boosted her onto his shoulders without breaking stride. Liban, pleased with the view, raised her arms and hooted so that all could admire her.

In 1976, Dr. Mary Leakey's team, working at Laetoli, made an astonishing discovery. They located a 3.5-million-year-old volcanic ash layer with preserved animal footprints. Dozens of different animals (including hominids) walked on the wet, muddy ash leaving a record of their journey. Extensive excavations at Laetoli uncovered more than 20,000 animal footprints, including those of elephants, rhinos, three-toed horses, giraffes, antelopes, a sabre-toothed cat, monkeys, birds, and even millipedes. The most astonishing discovery at Laetoli was the

thirty-foot-long trail of hominid footprints made by two different-sized individuals, walking side by side, heading north. At one point, something must have caught their attention: they stopped, turned, and looked. Lucy's feet were an identical match to the prints found at Laetoli.

This 3.5-million-year-old hominid print (below) is virtually identical to one a modern human would make. Most noticeable is the absence of a divergent big toe, which helps primates climb trees. Lucy had a reduced big toe like ours, which could not be extended and used for grasping and climbing.

Single Laetoli footprint

Laetoli hominid footprint trail

With his daughter seated firmly on his shoulders, Lorcan moved purposefully across the mud flats. Although smaller than Cano, he had become leader because of his inner qualities. He was, among the Family, the calmest male. His moves were decisive and seldom disputed, and then only effectively by Eba.

Yet Cano challenged him compulsively. He had often clashed with Lorcan since the death of Cano's brother, Broc, the previous leader of the Family. Cano's rebellions had always been checked by Lorcan's intensity. Lorcan's eyes, dark and set deep under his prominent brows, could catch and hold those of the Family and of animals longer than any of the others. He seldom made any sounds at all, except to bark sharply when leading or to hoot to raise an alarm. Although he was slightly shorter than Cano, his upright posture and measured assuredness made him appear larger.

Lorcan tempered his quiet intensity with a constant compassionate care for all the members of the Family. Beyond his personal attributes, Lorcan derived his authority from his strong bond with Lucy and Eba. Their insights and sensitivity to changes in the environment greatly influenced Lorcan's leadership. Unlike Cano, who attempted to gain rank through physical intimidation, Lorcan maintained his position by harnessing the abilities of every individual to benefit the group as a whole.

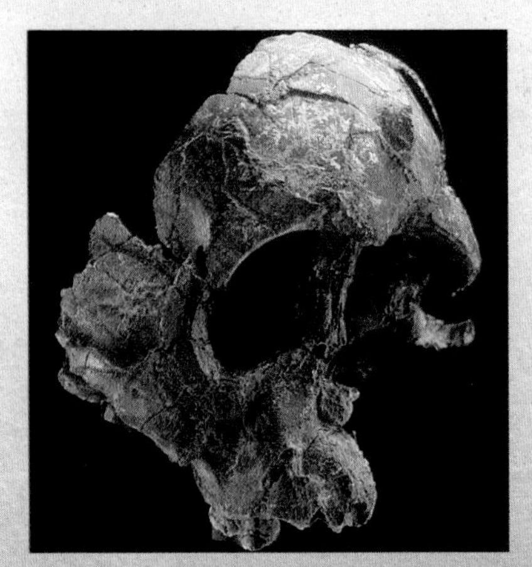

A. afarensis *infant cranium*

This cranium of an Australopithecus afarensis *infant was found at Hadar. Only the two milk (baby) molars are erupted. Features such as the strongly projecting face of adults are less developed in immature individuals.*

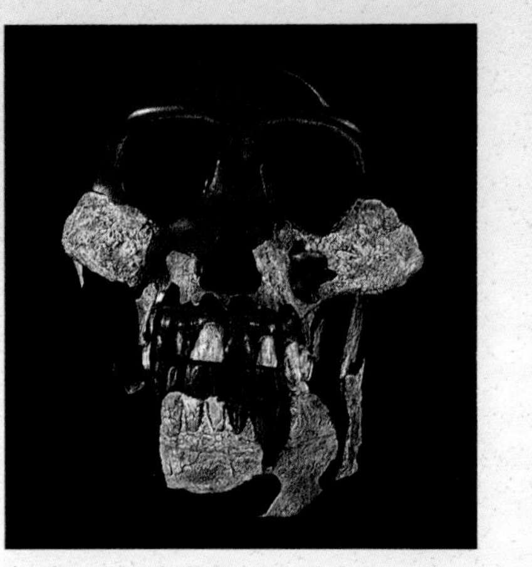

A. afarensis *skull, anterior view*

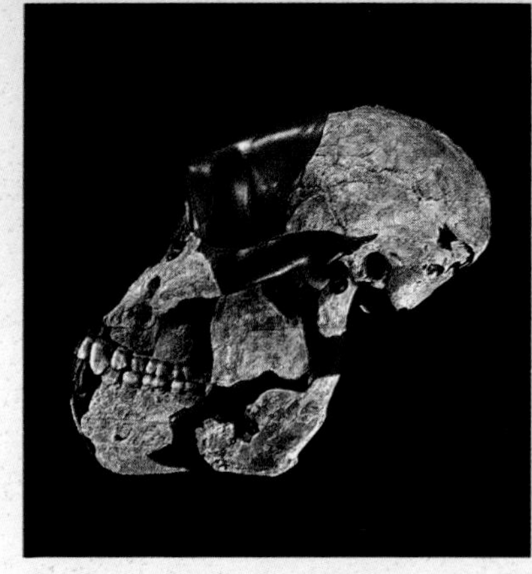

A. afarensis *skull, lateral view*

This reconstruction of a male Australopithecus afarensis *skull, fashioned by Drs. Tim D. White and William H. Kimbel, is apelike, with a large and projecting face resembling that of a chimpanzee. The cranium was small, enclosing a brain of only about 450 cubic centimeters, one-fourth the size of an average modern human brain.*

Roughened areas on the back of the skull indicate thick, strong neck muscles which balanced the skull on the spine. In order to move the large jaws and teeth, huge chewing muscles were necessary.

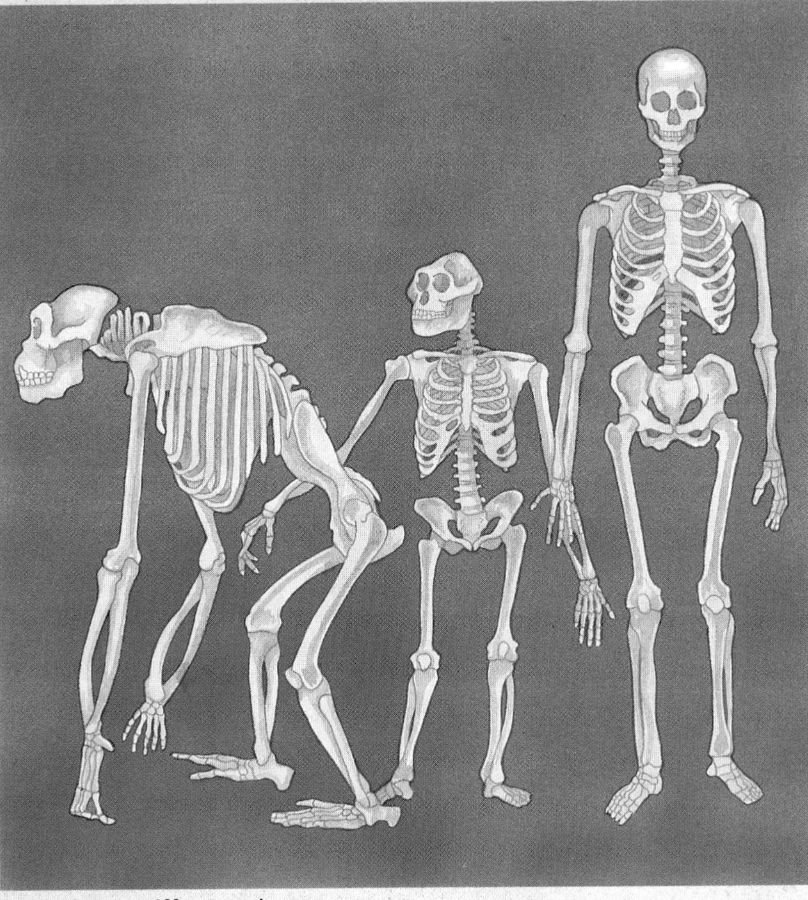

Chimp, Lucy, and Homo sapiens

Many differences between the skeletons of a chimp (left), a hominid like Lucy (center), and a modern human (right) are related to the way these animals walk. The skeletal plans of Lucy and the chimp are very similar, suggesting that Lucy's ancestors were quadrupedal.

One of the most intriguing problems facing anthropologists is trying to explain why our ancestors became bipedal. Such a dramatic change in locomotion must have brought with it some major advantages. Yet bipedalism is a slow and unstable form of locomotion.

One scientist hypothesizes that Australopithecus afarensis *was pair-bonded, that males and females mated for life. He postulates that bipedalism permitted the male to forage widely and gather food which could be carried back to be shared with his mate and offspring.*

At a small island of exposed grass in the sea of moist, gray ash, the Family paused to rest. Lucy lay down on the ground and went immediately to sleep with Lifi nursing at her breast. Looking down at his mate, Lorcan saw the exhaustion in her body and face. The fires and explosions, the premature birth, and the long trek had drained her.

Reaching down, he gently took Lifi out of her arms and handed the new baby to Eba. Then Lorcan began to groom his mate with infinite gentleness. His fingers ran softly through Lucy's fur, plucking off any small insects, stroking the matted places where the mud still clotted her hair. Then he rubbed the muscles of her lithe limbs. Lucy's eyes fluttered open for a moment to look at him, then shut again with a small sigh of pleasure as she drifted off into sleep.

Others of the Family were drowsing as well. Lorcan caught Dagan's eye and with a small sound made him understand that he should keep watch while the rest slept. As the adults rested, most of the children, excited by this strange, muddy landscape, chased each other in wide circles around the flats.

Lonnog, chasing Ciar, slipped on the smooth clay and into a small puddle. Pulling himself up, he thought he saw something move in the water. Curious, he bent over and peered in. It was his own face looking back at him. Startled, he brought his palm down onto the reflection, shattering it, and watched it form again as the ripples died away. As he pulled back, his hand pressed deep into the mud. Removing it, Lonnog saw the print his hand had left slowly filling with water. He pressed his hand down again and withdrew it. He looked at the two prints for a moment as water seeped in from the edges. He looked at his hand. For an instant, something sparked behind his eyes. He hooted softly and moved to put a third print between the first two. At that moment, his sister Liban scampered quietly up behind him and deposited a large gob of mud directly onto the middle of his back.

Our hands are similar to those of chimpanzees, but there are important differences. Chimpanzees walk on their knuckles and have shortened thumbs. They have a very powerful grasp, but they do not have the ability to touch the tips of their thumb and fingers together in a precision grip.

Some of the earliest hominid handprints can be traced to the Upper Paleolithic period of Europe, some 15,000 years ago. The handprint next to the spotted horse at the cave of Pech-Merle almost appears to be the signature of the artist who painted the horse. Handprints occur widely, as is seen in this dramatic picture of a Native American Anasazi hand. The stunning collection of hands in white and red is from the Cave of the Hands in Patagonia, at the southern tip of South America.

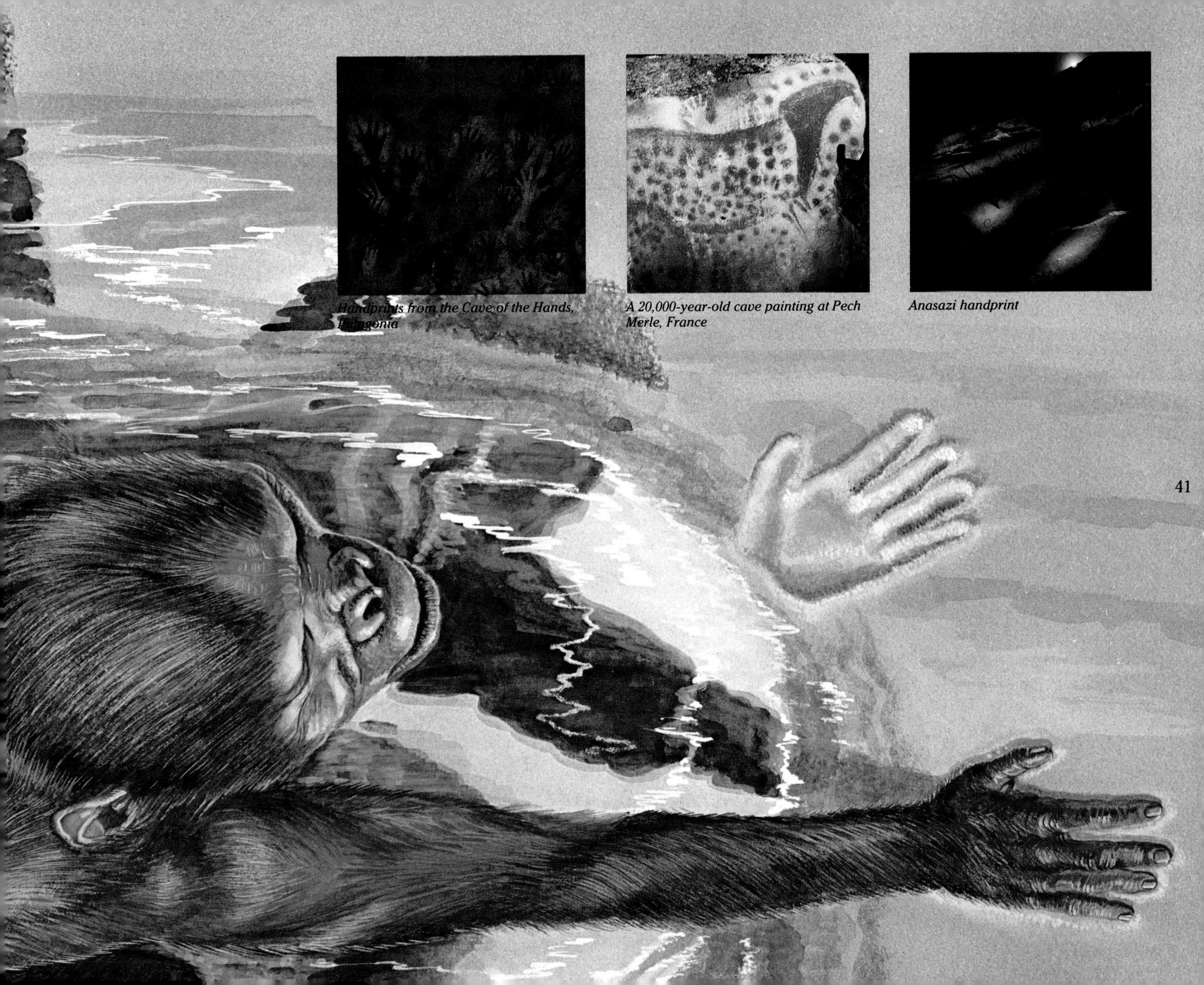

Handprints from the Cave of the Hands, Patagonia

A 20,000-year-old cave painting at Pech Merle, France

Anasazi handprint

The third handprint was forgotten as Lonnog whirled and, scooping up a large handful of mud, lunged squealing after Liban. Liban dodged nimbly to one side. Trying to slap the mud on her back, Lonnog missed his footing and, whooping, went sliding across the mud on his back.

Attracted by the shouts, the two five-year-old boys, Buan and Ocan, as well as Ocan's older sister, Ciar, sped across the mud to join the fun. Seeing Lonnog skidding across the slick surface, they quickly mimicked his technique. The mingled hoots quickly brought the rest of the children into the game. Soon all of them were encrusted with the gray clay, and large globs of mud flew through the air in every direction.

The children soon tired of flinging mud and rested in the sun. Liban ran up to Lonnog, who gave her a dismissive snort. But instead of running away she ran her fingers down his body leaving a pattern of lines in the dried mud on his fur. Then she reached up to the mud on his cheeks and did the same thing. The other children stared, finding this as novel as the mud fights, and began to mark each other's bodies and faces in the same way. As soon as they all had created patterns to their liking, they celebrated by running off into the mud flats and exuberantly sliding in the moist ash all over again. Bran and Anu, the oldest youths, began to lead the rest away from the adults and toward the river, but Dagan barked out a sharp warning. Predators were still active in the lowlands.

In virtually all cultures people decorate and paint their bodies in one way or another. These images from New Guinea, Australia, and Africa demonstrate how elaborate and dramatic the results can be. Individuals may think that such decorations make them more attractive. Specific decorations may convey the social status of the wearer. Particular designs are associated with special ceremonies such as marriages, funerals, puberty rites, and other occasions. Such adornment often endows the person with a more powerful image. Perhaps seeing coloration, as in this type of baboon called a mandrill, people were inspired to decorate themselves.

New Guinea highlands man

Australian Aboriginal children

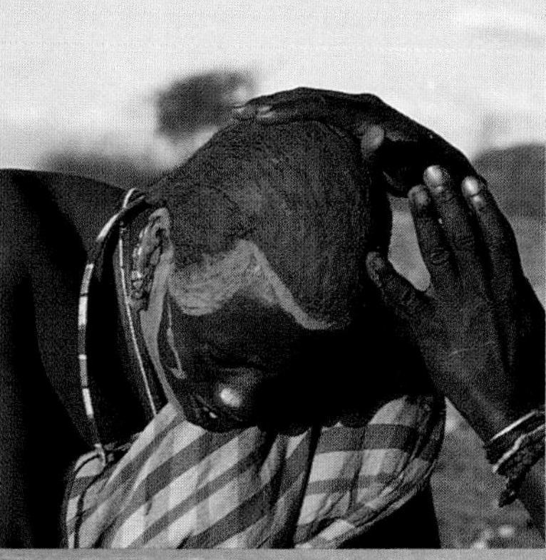

Masai, mud in hair

Mandrill

Wodaabe male dancer

New Guinea Mendi girl

A large portion of the grove edging the stream had been ignited by a flow of lava in the night. The rain had extinguished it for the most part, leaving blackened earth and pillars of twisted, burned trunks. Where the lava cooled, pools of water steamed and covered the ground with a thick, white fog.

As they passed a small knoll, Lonnog hunched down and made a soft sound. Going over to him, the adults found a pile of unburned seed cases that had been cracked open by the heat. In the midst of the seed cases were the remains of a bird's nest. The eggs had been split open and cooked by the fires, but remained unburned. In the same manner, the seeds inside the cases were toasted but, as Lonnog and the others soon found, quite edible. The eggs, although strange to the Family, who always ate them raw whenever they could find them, were equally tasty in their cooked state.

They traveled through the charred grove and out onto the hardened field of black lava, fog clinging close to its surface at first but gradually fading away. Some places on the lava flow were hot to the feet, others merely warm. Lorcan noticed that Eba was still favoring her leg and offered to help her, but she would have none of it. Liban, upset by the heat of the stone, took her place again on her father's shoulders. Around them, the complete absence of life made the entire Family nervous. The night before, this land had been the blood of the earth; today it was steaming rock.

Fire has always been an important factor in the Serengeti. Most fires are ignited by occasional lightning strikes or volcanic eruptions. In the past, it was not unusual to see more than seventy-five percent of the Serengeti Park burn. No wonder some trees have become fire resistant, developing thick bark or a living root underground.

When fire and elephants destroy old trees, more grass grows. Grazing antelopes, by removing the grass, reduce the threat of fire. Over a long period of time, trees might grow again, changing the landscape to more of a woodland, enticing the elephants to begin the cycle over again.

Fire on the Serengeti

While wandering across the fog-enshrouded lava field, Bran suddenly hooted loudly, attracting the others.

It was the scorched skeleton of a dinothere that had been overtaken by the lava. Covered with a thin shell of red-hot ash and rock, its flesh had burst into flame and been utterly consumed, leaving only the bones nestled in the cooled lava like an exhibit in a frame. The Family looked at the strange apparition from every angle and then began to probe at it gingerly with sticks.

Locan noticed that the crater in which the skeleton lay conformed to the profile of the beast when it had been alive. He gestured at the outline of its legs, head, and curled trunk. The Family became even more uneasy. They stepped back and looked at each other with wide, nervous grins. Some of the children began to scream in a frightened way until Lorcan silenced them with a grunt.

Then he turned and led them off the black stone river and away from the strange, dead beast, toward the higher grasslands and the river beyond.

Volcanic eruptions may preserve impressions of living organisms, like the unique footprint trail at Laetoli. Another well-known example is Pompeii, where a glowing avalanche of volcanic ash buried the town and its inhabitants. Archeologists poured cement into the cavities (molds) formed by bodies and made casts of the people and animals of Pompeii.

The elephant skeleton pictured here is enclosed in a lava flow in eastern Zaïre. This skeleton, because it was not covered by water or volcanic ash, will not be preserved in the geological record.

Casts of humans at Pompeii

Cadaver of an elephant asphyxiated by volcanic gases in eastern Zaïre

The sun was high in the sky when the Family reached the end of the sloping lands and stood on a ridge that fell away to the river below. Looking down, Lorcan gestured with his stick, and the Family crouched in the grass and became still.

As they watched, a herd of dinotheres emerged from a thick stand of trees on the far side of the river and paused on the bank. Rapids churned through the deep brown water. A large bull tossed its ponderous head and raised its trunk into the air, then waded out into the stream toward the opposite bank. The herd followed reluctantly behind him.

Bringing up the rear, a small calf waded in just behind its mother. Losing its footing as the river deepened, the calf was swept downstream by the current, trumpeting plaintively. The mother lunged after it, seeking to place her body between the calf and the current. With the mother's trunk wrapped around it, the calf struggled toward the bank. The current swept them along. By the time they reached the opposite side, the bank had become too steep for either the calf or the mother to climb. Exhausted, the calf slipped back into the river and was swept farther downstream. At times only the tip of its trunk was visible above the thick, brown surface of the river. Its mother, too, was held in the water's grip, swept along after her child toward the lake in the distance.

The Family now had no doubt that it was dangerous to cross the river at that point. Rising up from the grass, they moved quickly along the ridge toward the broader deltas where the river widened, providing a safer crossing.

Elephant swimming

Due to the enormous surface area of its skin, the Dinotherium *probably suffered from severe water loss through evaporation in the torrid African sun. Elephants today cover themselves with mud or wallow in streams, where they may drink over 50 gallons of water daily to minimize and compensate for these losses. Despite their bulk, elephants are excellent swimmers, using their trunks like snorkels for breathing while their bodies are submerged. Dinotheres probably forded rivers in similar fashion to their present-day relatives.*

When they arrived at the flatlands they found shallow water and new bars made from sand, soil, and ash. Large stretches of the river had been cut off from the main channel. In one shallow pond, a lava flow had made the water hot enough to kill all the fish that had been trapped there. Now, in the midday sun, the pond was rife with the bodies of fish floating in the pool. As the Family rushed to this unexpected feast, a cloud of birds eating the fish on the banks rose as one, only to settle back down once they had passed.

The Family was busy collecting and eating the fish when the surface of the pool suddenly erupted, spraying water all around. A crocodile had found the pool earlier in the day and claimed it as his private preserve. The limitless supply of food and the high temperature of the water had driven the huge reptile into a frenzy. It slithered through the water, snapping chunks off some fish and swallowing others whole. Each fish, although dead, was carefully stalked by the croc. It crept up slowly and lunged violently at each dead fish to make certain it could not escape. After one charge, the croc turned its head, eying the Family as if considering a change of diet. Then, as if shrugging, it turned away and continued to feed.

Detailed study of the teeth of Australopithecus afarensis *fossils indicates that these early hominids were essentially vegetarians. They would have eaten seeds, nuts, berries, fruit, roots, tubers, and flowers. Lacking piercing canines or slicing back teeth, it is unlikely they were carnivorous. Being opportunistic omnivores, they must have occasionally added some protein to their diet. They probably ate larvae, insects of various kinds, lizards, snakes, and perhaps fish. Although* A. afarensis *undoubtedly ranged over large areas in search of food, they probably were not very effective meat scavengers. Even if they were able to drive a predator from its kill, they would have had a difficult time eating meat without stone tools to cut it with. Also, they would have faced fierce competition from scavengers better equipped to profit from an abandoned kill.*

As the floodwaters receded the bodies of wildebeest and hipparion horses were half-buried beneath the silt or fully exposed like heaps of wood just above the waterline. The corpses attracted many of the surviving predators and scavengers from miles around to the water's edge.

With such a rare abundance of food laid out under the sun, the need to hunt and kill was no longer necessary. Instead, predators and scavengers came to the banquet in waves, each according to their station, as if some ancient seating order were being rigidly observed. Large cats and packs of hyenas were the first to feast. When they were sated they loped off to sleep in their lairs or under the shade of nearby trees. Then small jackals, river rats, and other rodents emerged from all directions to nuzzle and feed in the rich mounds of carrion. At the same time, flies and beetles swarmed over the food.

Overhead, the afternoon sky darkened as large flocks of vultures spiraled down to the banks. Solitary eagles swooped down and, along with the storks and other birds, tore gobbets of flesh loose from the carcasses and gulped them down.

The Family, sated on fish, walked past the remains of a wildebeest. So abundant was the food along the banks that this animal was still virtually intact. Small white-headed, white-backed, and Rüppell's vultures had been worrying the carcass for some time, but their beaks were too weak and ill-adapted to the task of opening the thick hide. Instead, they hopped about the corpse, squabbling among themselves. Suddenly, a huge lappet-faced vulture dropped on the wildebeest like a stone. The squabbling birds stepped back and waited as its large beak lanced down and, with a jerk of the bird's head, tore a gaping hole in the belly of the animal. Jerking again it opened the entire underside of the body. In a flash of whirring feathers, the carcass was buried beneath the hunkered bodies of the vultures. The bald-headed ones probed deep within the belly, seeking the soft inner organs, while the others moved along the outside, stripping away the skin to get at the lean muscles.

The various types of vultures on the Serengeti exhibit different scavenging behaviors. Each species arrives at the site of a kill at a different time and specializes in a different part of the carcass.

The lappet-faced vultures are powerful scavengers, able to tear off tough and sinewy meat. The hooded and Egyptian vultures rip the smaller pieces of meat from bones with their sharp beaks. The white-backed and Rüppell's vultures have bare necks, allowing them to enter an animal carcass and pull out large, soft morsels.

Under the trees the megantereons lolled peacefully. A female, roused by the heat and the food, gave a soft roar and beckoned a male to her side. The male, not certain at first of her acceptance, circled cautiously. She moved about to present herself. The male mounted her with a lunge and, within moments, moaned. He withdrew and, rolling to one side, his mate gave him a soft pat on his jowls with a languid sweep of her massive paw. They lay side by side in the grass, and after a time the ritual was reenacted, then again, and again.

Wading in the shallows close to the bank, the Family passed close by the pride. They knew that they had nothing to fear from the predators. The big cats never killed except to satisfy their hunger. Once fed, they were content to laze their time away sleeping, preening, and playing with their cubs in the shade.

Lucy paused with Lifi at her breast, as two of the cats and a cub lapped water from the lake. One of the cats, a female, gripped her cub by the nape of its neck and carried it back to the pride's nest. The mother held the bobbing cub with her incisors, careful not to hurt it with her daggerlike fangs.

Male and female lions with zebra

The other cat raised its head and gazed at the strange group splashing past it in the water. It calmly lowered its head and went on lapping up the murky green water as all the members of the Family passed by. Even the children made no sounds. The cats, they knew, had no interest in them. Still, they were to be respected.

Lions are sexually mature at about two years. Prior to mating the male and female lions may rub heads and sniff each other. Mating activities can continue for up to six or seven days, during which time the lions do not eat. The female initiates the encounter, prompting the male to mount her every fifteen to twenty minutes. After three and a half months a litter of one to five cubs is born.

Hungry predators move in a characteristic way that frightens their prey. After a successful kill, the predators are obviously satisfied, relieving the fright among the prey. Harmony quickly replaces fear.

Down the river, trees grew close and thick. Probing beyond the trees, the Family found a small glade. As they emerged from the bush, they encountered a small troop of gelada baboons who had wandered in from the opposite direction. Both groups froze for a long moment and carefully took each other's measure. Neither enemies nor allies, the Family and the baboons silently regarded each other across the clearing.

Then, as is usual when two groups like this meet, the younger children were the first to break the impasse, playing chasing games. The older children remained in a cluster near the adults.

A small baboon, younger than the others, seemed to take an immediate fancy to Liban and presented his head to her for a scratching. Liban was delighted. Soon she and the small baboon scampered about the glade in ever-widening circles, until they fell to the ground, wrestling.

At first, though Liban was larger, the young baboon seemed to have the advantage. When he drew back after a rolling tussle, Liban snatched up a large frond and, waving it about, began to chase the young baboon around the glade with it. The sight was too much for the baboon. His excited barking mingled with Liban's high chittering laughter, and much of the tension in the glade dissipated.

Lucy was pleased for Liban, but knew how quickly play could turn into something more serious. If Lucy were to interfere, the baboons might attack, believing that she was trying to harm their child. Likewise, any interference from the adult baboons might provoke a similar response from the Family. These concerns did not trouble Liban or her new playmate. They ran about the glade, then disappeared into the brush on the far side. After a few moments, Lonnog, protective of his little sister, followed the pair through the undergrowth.

Jane Goodall has observed numerous instances in which olive baboons and chimpanzees have interacted. Young chimps and young baboons will often play with one another. Sometimes the play becomes very excited and even aggressive. The chimp may stamp the ground and throw rocks and sticks, driving off a baboon. At other times the baboon gets the upper hand and drives off the chimp. If the play becomes too aggressive, with animals hitting or biting each other, female or even adult male chimps will interfere, defending the offspring.

Chimpanzee threatening a baboon

Bursting out of the dense brush, Lonnog saw Liban and the baboon some distance out on the open savanna hovering over something in the grass. The boy ran over to them and saw that they had discovered an unguarded clutch of ostrich eggs. At first, the baboon attempted to lift one of the huge eggs, only to have it slip from his hands. The baboon rubbed his hands in the dirt to coat them with grit. He lifted one of the eggs successfully, chittered with pleasure, and began to lope back toward the glade. Liban rubbed her hands in the dirt, grabbed an egg, and set off after the baboon. Lonnog followed them both for a distance to make sure they returned safely, then he turned back toward the nest. While he had left it, however, an Egyptian vulture had landed next to the nest. It picked up a stone in its beak and threw it at the thick shell of an egg. Lonnog leapt forward, waving his arms, and the scavenger flapped away.

Alone, Lonnog decided to eat an egg on the spot. Taking his lead from the vulture, Lonnog knelt on the ground, picked up

The Egyptian vulture is noted for an exceptional behavior, which allows it to benefit from the protein inside an ostrich egg: Using its beak, not designed to break the thick egg shell, the vulture picks up a stone and throws it at the egg.

Egyptian vultures breaking ostrich egg with stone

a stone, and began to hammer his way through the thick shell. Slowly, he managed to chip a small hole in it. His knees felt a rumbling in the ground. Whirling around, he saw the immense bulk of an angry brown female ostrich bearing down on him across the plain.

Dropping the egg, Lonnog sprinted for the closest tree. The outraged ostrich ran past the nest to pursue the boy, its long neck stretched out in anger at the thief. Shrieking, Lonnog spun about, swung his fist, and struck the small head of the bird, stunning it for a moment. Sprinting under the bird's extended neck, he leapt into an acacia tree and scrambled up through its thorns, tearing his flesh in a dozen places, screaming at the top of his lungs.

The only flightless bird native to Africa is the gangling ostrich, the world's largest bird. Ostriches have long, powerful legs, which help them run quickly (up to thirty-five miles per hour), and are also powerful weapons.

The males, who defend the territory, are black and white in color. The male courts and mates several females. More than one female may lay eggs in a single nest, a simple hollow in the ground that can contain more than one hundred eggs. Hens lay eggs in more than one nest. The mixture of eggs in one nest assures a high degree of genetic variation in the offspring. If one nest is destroyed by scavengers, some eggs laid by the same bird will survive in nests elsewhere.

Hens lay up to thirty eggs, not all of which produce a chick. The male ostrich sits on the eggs at night, the female during the day. Presumably, the female—lighter in color—is less conspicuous in the daylight.

Male and female ostriches with eggs

Lorcan and Dagan came running through the brush. On hearing Lonnog's terrified screams they dashed forward without a second's hesitation. Lorcan brandished a large stick while Dagan, scanning the ground, had to settle for a smaller one.

As they closed on the nest, they saw Lonnog struggling higher into the thorn tree while the ostrich jabbed at the boy with her hard beak. Lorcan and Dagan split forces and, shouting, rushed the bird with sticks waving and teeth bared, driving her away from the tree. Lorcan turned to the tree, while Dagan jumped back and forth around the enraged bird, keeping her at bay.

Lonnog clambered higher in the tree and clutched the branches tightly. Lorcan, standing at the foot of the trunk, shook on the branches and gestured impatiently for the boy to come down. Lonnog, looking out across the plain, saw the ostrich's mate coming over the grass at high speed. He froze on the trunk until Lorcan's stick landed on his foot with emphasis. Lonnog plummeted to the ground, landing on his feet.

Lonnog began to run to the safety of the thick underbrush, only to find his arm grabbed by Lorcan as they started toward the nest. Unguarded ostrich eggs were a rare and rich prize on the savanna. Lonnog hesitated. Then, seeing his father move toward the nest, he followed. While Dagan beat the female back with his stick, Lonnog and Lorcan raced to the nest. The black male ostrich bore down on them with increasing speed, its head and neck stuck out before it like a spear, its calls high and screeching.

Lonnog grabbed two eggs, then spun and dashed for the brush. He was vaguely aware of his father, Lorcan, holding two large eggs close to his chest, running for the safety of the glade. The grass spread out in a blur of yellow ochre. The thundering footsteps of the ostrich grew nearer. Lonnog gained the first fringe of the brush, pitched the thick-shelled eggs before him, rolled into a clump of tall yellow grass, and became as still as a stone.

The rapid steps of the giant bird closed on him, passed, and faded through the brush. Then they returned, moving this way and that. Gradually all fell silent. Lonnog took one breath, then another. He froze again as a twig snapped nearby. His shoulder was suddenly pushed hard and he wailed and rolled over . . . into a storm of chattering laughter.

Lorcan and Dagan loomed above him jumping up and down, their eggs clutched close to their chests. Lonnog struggled to his feet and parted the grass to look out on the savanna. Both ostriches were standing over their nest, guarding the ten eggs that remained.

KOF

After returning to the clearing and the excited Family, Lorcan raised one egg to his ear and shook it. He put it down and did the same with the next. He placed it with the first. He shook the third, looked at it closely, and shook again. This egg he placed to the other side.

Lonnog followed his father's actions curiously. He picked up the first egg and shook it—it made no sound. Then the second—again silence. He reached around his father and shook the third egg. From this one came a faint thumping sound. His father added another egg that also felt as if something was moving within. Lorcan moved about the glade, looking down into the short green grass. He stooped and quickly dug a long thin rock out of the damp soil.

The !Kung San of the Kalahari Desert in southern Africa break a small hole in one end of an ostrich egg, fill it with water, and use it like a canteen. They bury several eggs filled with water at strategic places in the desert and use them as an oasis.

!Kung San woman filling ostrich egg with water

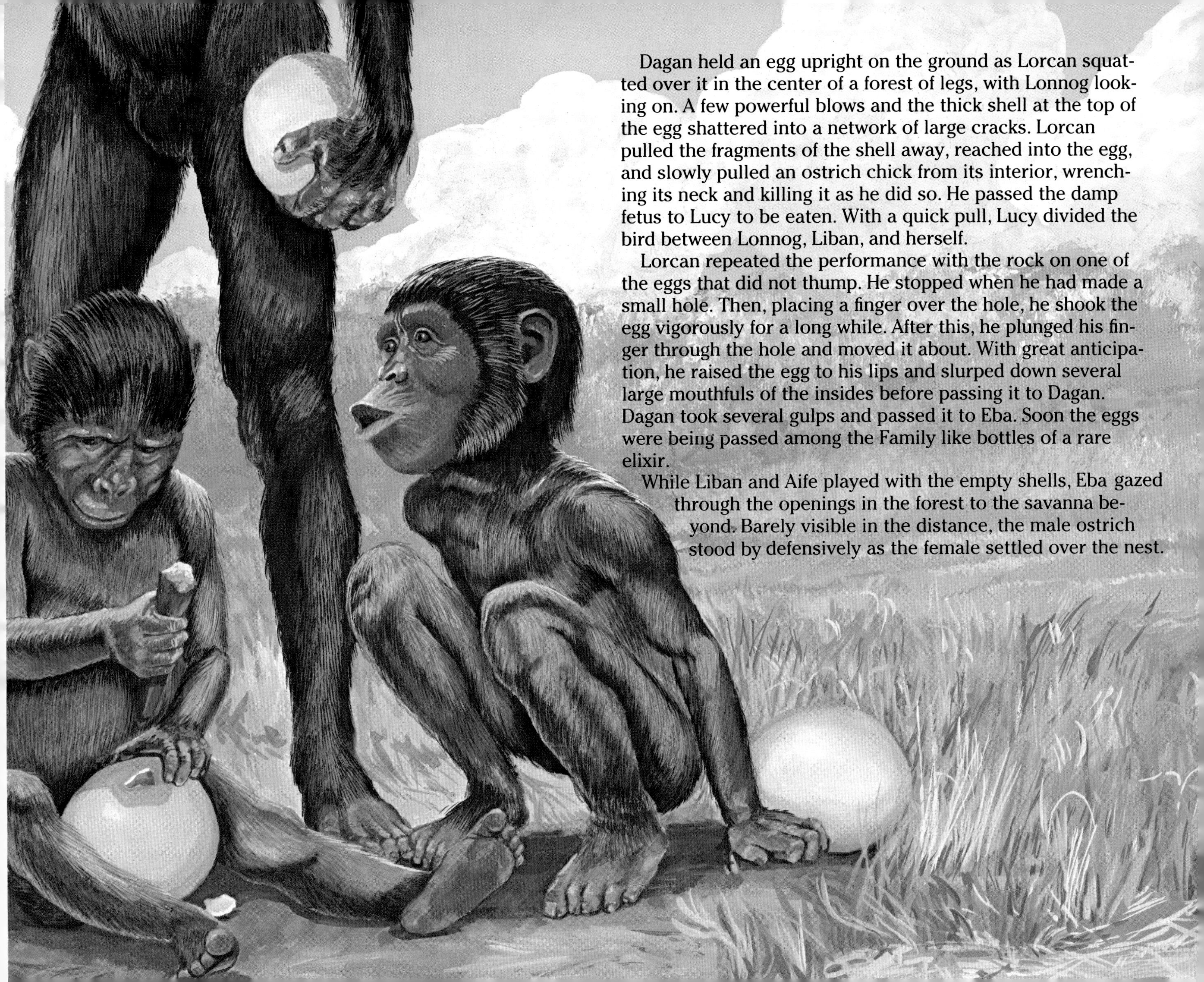

Dagan held an egg upright on the ground as Lorcan squatted over it in the center of a forest of legs, with Lonnog looking on. A few powerful blows and the thick shell at the top of the egg shattered into a network of large cracks. Lorcan pulled the fragments of the shell away, reached into the egg, and slowly pulled an ostrich chick from its interior, wrenching its neck and killing it as he did so. He passed the damp fetus to Lucy to be eaten. With a quick pull, Lucy divided the bird between Lonnog, Liban, and herself.

Lorcan repeated the performance with the rock on one of the eggs that did not thump. He stopped when he had made a small hole. Then, placing a finger over the hole, he shook the egg vigorously for a long while. After this, he plunged his finger through the hole and moved it about. With great anticipation, he raised the egg to his lips and slurped down several large mouthfuls of the insides before passing it to Dagan. Dagan took several gulps and passed it to Eba. Soon the eggs were being passed among the Family like bottles of a rare elixir.

While Liban and Aife played with the empty shells, Eba gazed through the openings in the forest to the savanna beyond. Barely visible in the distance, the male ostrich stood by defensively as the female settled over the nest.

At midday the equatorial sun was directly overhead, and the grove was glimmering with light while the Family still lolled in the shade. Lorcan at last roused himself from where he had been resting with Lucy and Lifi, and the rest of the Family stirred as well. Soon they were again moving across the savanna.

By this time the sun had evaporated all the remaining puddles from the late-night rain. Clusters of small white clouds dotted the sky, and the continual chant of weaverbirds rang out over the plains. The Family strolled between calm herds of antelope and gazelle that were also moving to the river to drink. This was the safest time for animals to go to the water, since most predators preferred to nap during the heat of the day. As the Family moved along they gave hardly a thought to the leopard dangling in a tree or the hyena pack that trotted well ahead, on its way to its lair somewhere along the eroded banks of the river.

Farther along, the Family skirted a pride of homotheres inside an acacia grove. The adults were resting while the cubs, never still, chased each other like spotted balls of furred energy. As the Family looked on, a male was awakened by two scuffling cubs and scattered them with one languid sweep of his paw.

Eba shuddered at this movement. It made her recall the terrible night when her mate, Enan, had been killed. They had all been sleeping, but were awakened by Enan's screams as the great cats' saw-edged fangs cut like scimitars through his flesh, shook him to death, and dragged his body off into the darkness in the space of a few awful seconds. Eba moved a little quicker than usual, not wanting to be left behind at this point.

Some Australopithecus *fossils found in South African cave sites may be the remains of leopard kills. One scientist believes that leopards carried hominids into trees near cave openings and, as they devoured them, portions of the bodies fell into the sinkholes. In one* Australopithecus *skull, there are two holes that correspond exactly to the distance between the lower canines of a fossil leopard jaw.*

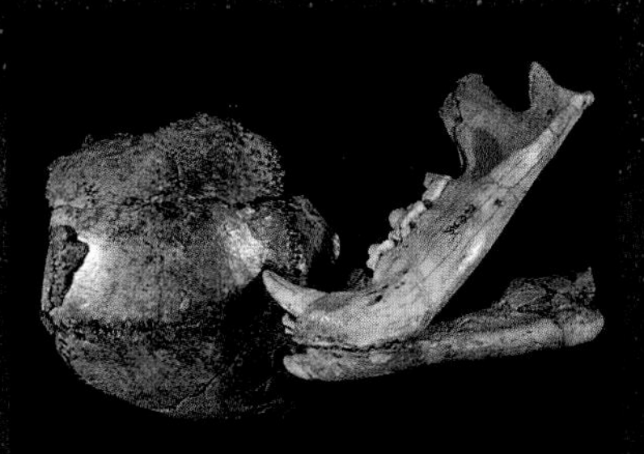

Fossil leopard jaw with Australopithecus *skull*

Homotherium *was a cat, about the size of a lion, with shorter, more curved canines than its contemporary,* Megantereon. *The canines possessed a crenulated (serrated) edge. Sometimes called the scimitar-tooth cat, it is well known from excavations in Eurasia and North America. It was a curious-looking animal with greatly elongated forelimbs and short hind legs.*

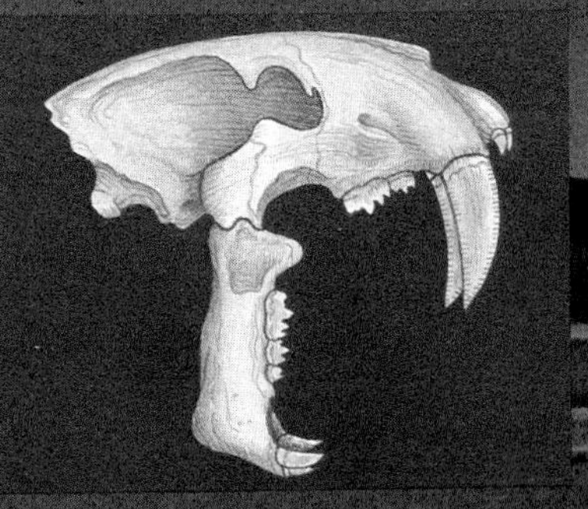

Homotherium *skull*

In carnivorous animals, specialized shearing, or carnassial, teeth are very effective for slicing meat. In sharks, Homotherium, *and even the* Tyrannosaurus *some of the teeth have a serrated margin, forming a specialized cutting edge, like a bread knife.*

Serrated teeth of great white shark

Serrated teeth of tyrannosaurus

Lucy, with Lifi riding contentedly over her shoulder, strode through the grass of the plains, following the others. Far off, she saw the noon heat shimmer along the horizon. As she watched, a lake seemed to appear out of nowhere. Running ahead of Lucy like a small scout, Liban turned and hooted high and clear at her mother, gesturing toward the vast expanse of water that had appeared so mysteriously.

Lucy's free arm swept out and away from the lake, directing her daughter to ignore it. Lucy had seen the lake many times before. She had even, during the dry season when water was scarce, run toward it, but no water was ever found.

Liban turned from the shifting mirage and spied a small monkey exploring a mound of dinothere dung for bugs hiding within. As she approched, the monkey scampered off into the grass. She started to follow it, but her attention was grabbed by a scuffle to one side of the odorous mound. A pair of dung beetles were struggling with a small ball of excrement. The girl squatted down and watched the brown beetles pushing and pulling this way and that, seeming to get nowhere with the clot. Slowly, inch by inch, the beetles moved it closer to their burrow. Curious, Liban reached out with her hand and with a deft flick of her forefinger knocked the ball away from the beetles.

Liban was amused to see them scamper over to the ball and begin to move it back to the burrow again. After some moments, they got their prize to the entrance.

A quick click came from Lucy. Liban turned from her play to follow her mother, leaving both the ball and the beetles unharmed.

Elephants play an important role in the survival of the very acacia trees they eat and destroy. A parasitic beetle often lays its eggs in the seed pods of the acacia tree. The larvae eat and destroy up to ninety-five percent of the seeds. If elephants ingest the infected pods, their digestive juices kill the beetle larvae. The resulting healthy seeds in the elephant dung are buried by the dung beetles, ensuring the birth of a new acacia tree.

These wildebeest appear to be standing in a lake, but it is only a mirage. A mirage is an optical illusion caused by light rays being bent (refracted) through layers of air of different temperatures and densities. Such mirages are curious, because as one approaches the apparent lake, it seems to recede; and as one walks away, it appears to follow.

African mirage

On the Serengeti, there are a number of organisms which play a special role in consuming dead animal and plant matter. These creatures break down organic remains and release their nutrients back into the soil.

These decomposers include vultures, flies, crickets, bacteria, and the dung beetle. Dung beetles may descend on a two-hundred-pound pile of elephant dung and form it into small balls which will be rolled away and buried up to three feet underground. The beetle deposits eggs into the dung balls and, when the eggs hatch, the larvae feed on the dung.

The scarab was sacred to the ancient Egyptians, who compared the dung ball to the sun. They believed that the sun was pushed across the sky by a giant dung beetle. The scarab was considered to be the giver of life.

Scarab at Egyptian Museum

The Family was working its way back into the grove by the river when they saw a strange shape in the low branches of a tree. A pair of hammerkop birds had constructed an enormous nest. The hammerkops built their nests from twigs, straw, and mud and decorated them with choice items from the river, the savanna, and the forest. The result was a hodgepodge of branches, feathers of all kinds and colors, bones, antelope horns, pig tusks, dangling snake skins, porcupine quills, and dozens of other objects.

Everything else in the valley related to the other things in a natural harmony that the Family understood. This strange assemblage of unrelated items definitely seemed unnatural and struck fear into the Family. Eba refused to walk beneath the huge nest. She sat down while the others chattered around her.

The two black hammerkops emerged from a hole in the nest above the Family. Oblivious to their observers, the birds began jumping on and off each other's backs, bobbing from side to side. The mating ritual, combined with the odd nest, was too much for the Family and they withdrew back into the brush to look for other ways down the river.

Lorcan led the Family along a sandbar at the river's edge. The shore they passed was covered with mangled skeletons and broken bones. Small rodents, flies, and black beetles scuttled around them.

As they made their way to the shore through the field of bones, Lonnog yelped sharply and sat down quickly. While the rest of the Family watched, he pulled a long sliver of razor-sharp bone from the bottom of his foot and held it up. He looked first at his foot and the small gash that the bone had made in it. Then Lonnog ran his thumb along the edge, slicing his finger. Growling, he thrust the bone deep into the sand and hobbled after the Family.

Hammerkops are best known for their hundred-pound dome-shaped nests. The remarkably strong nest may take up to six months to complete. In addition to sticks, reeds, and grass, these birds use bones, plastic, paper, cloth, and other rubbish as their building materials.

Hammerkop nest

Experimental use of a bone digging tool

No bone tools have been found with Lucy. Polished bone tools—which may have been used to dig up roots and tubers —are known from fossil sites in South Africa nearly 2 million years in age. Experimental work with bone tools (shown here) demonstrates how effective they can be for digging up roots, which are a good source of food.

Around the bend, the river widened. A long, shallow marsh stretched off to one side. The adults of the Family lolled in the shade of ebony trees and listlessly groomed one another, while the children—with lots of whooping and chattering—began to run back and forth in the shallows.

The water, warm with the heat of the sun, was thick with lily pads and other river plants. One of the children bent down and took an entire plant from the sandy bed, making a quick snack of it. When Lonnog bent over to do the same, Liban stormed up behind her brother and knocked him into the water.

Lonnog emerged hooting and waving his hands, the thin fur on his chest dripping water and shreds of vegetation. He set off after his sister. The other children joined the pursuit with excited grunts. The children circled out from Lonnog, forcing Liban deeper into the water.

Lucy and Lorcan sat together on the shore watching over the squealing children as Cano, Cera, Dagan, Brig, Aine, and old Eba drowsed. Suddenly, Lucy noticed that one clump of vegetation was moving against the current. She looked more carefully. A prowling crocodile was slithering low in the water toward the children. Lucy shrieked a warning.

The children did not even look around to see what the danger was. To them, the warning cry meant only one thing—to return to the safety of the group as quickly as possible. In seconds, all the children had reached the waiting adults. The croc closed in but was met by a hail of rocks from the snarling Family. A few sharp blows on its head convinced the reptile that easier prey was to be had elsewhere. Diving beneath the water, only the long receding ripples betrayed its slow retreat.

Binocular (two-eyed) vision permits stereovision, allowing good depth perception. In addition to being useful in the trees, stereovision is important when throwing objects, from simple rocks to sophisticated hunting weapons. Stereovision may have developed in early primates in response to eye-hand coordination being important for catching insects.

After their encounter with the crocodile, the Family proceeded farther downstream and crossed a broad, shallow portion of the delta, dotted with sandbars and small wooded islands. The adults carried the youngest children on their shoulders while the older youths clung to their parents for safety in the current.

Lucy rested on the shore of a larger island with Lifi and the smaller children while the others foraged. Eba led the women and older children while Lonnog joined his father and Dagan. Cano preferred to forage alone.

Lorcan and Dagan waded into the green water and stood perfectly still with sticks held high over the surface. Their eyes slowly scanned the surface of the river. Standing motionless in the shadows of the trees at the edge of the pool, a white egret darted its head forward and speared a small catfish, flipped it into the air, and swallowed it. Encouraged, the men spotted ripples on the surface and also struck, unsuccessfully. The glare of the sun directly overhead obscured their targets below the water's surface.

African storks, herons, and egrets have a varied diet, which includes frogs, insects, worms, and snails; some are particularly fond of fish. The yellow-billed stork, pictured here, is enjoying a catfish. By watching these birds, our ancestors may have been inspired to try their hand at catching fish.

Yellow-billed stork with fish

Lonnog spotted a wood stork fishing in the sun. The tall bird hunched over with its head hanging in the shadow of its outstretched brown wing. Within moments it snapped a small fish within its bill. Inspired, the boy bent low over the water with his hands outstretched just at its surface. After moments of frozen anticipation, he could clearly see a large catfish as it entered the long shadow cast by his body. His muscles tensed as the unaware fish glided just below his fingers.

With a sharp cry Lonnog plunged his hands into the water and snatched the fish out. Its slimy green body twisted and wriggled and slipped through his fingers to splash back into the water. Quickly, Lonnog grabbed it once more. Again it slipped out and regained the river. With a growl, the boy plunged out of sight beneath the water. In a few seconds he came spouting to the surface, his head and shoulders covered with rotting river weed, but with the fish caught triumphantly in his hands. The sight of the boy decked out in weeds and waving a fish about was too much for Dagan and Lorcan. They ran laughing back to the bank. Lonnog didn't mind their laughter. He had the fish—*his* fish.

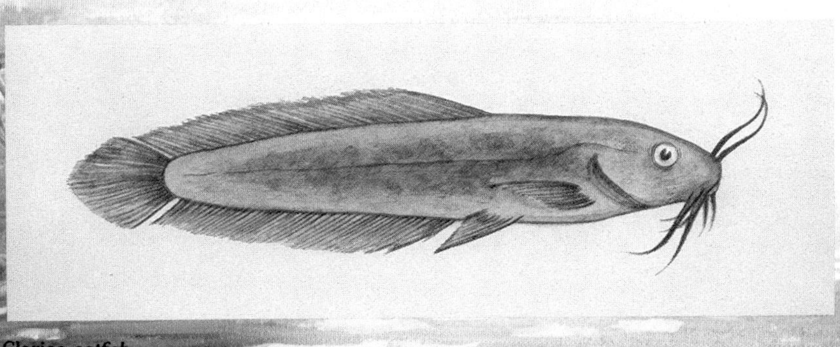

Clarias *catfish*

While others pursued fish, Liban lay in a small grove of papyrus near the calm water. Water spiders skittered along the surface on endless patrols. A small python glided below the lilies while bright green frogs squatted on the leaves, springing away when the python approached too closely and disappearing into the water with sharp, plonking splashes. The ripples drifted out, vanishing among the shoots and branches at the shoreline.

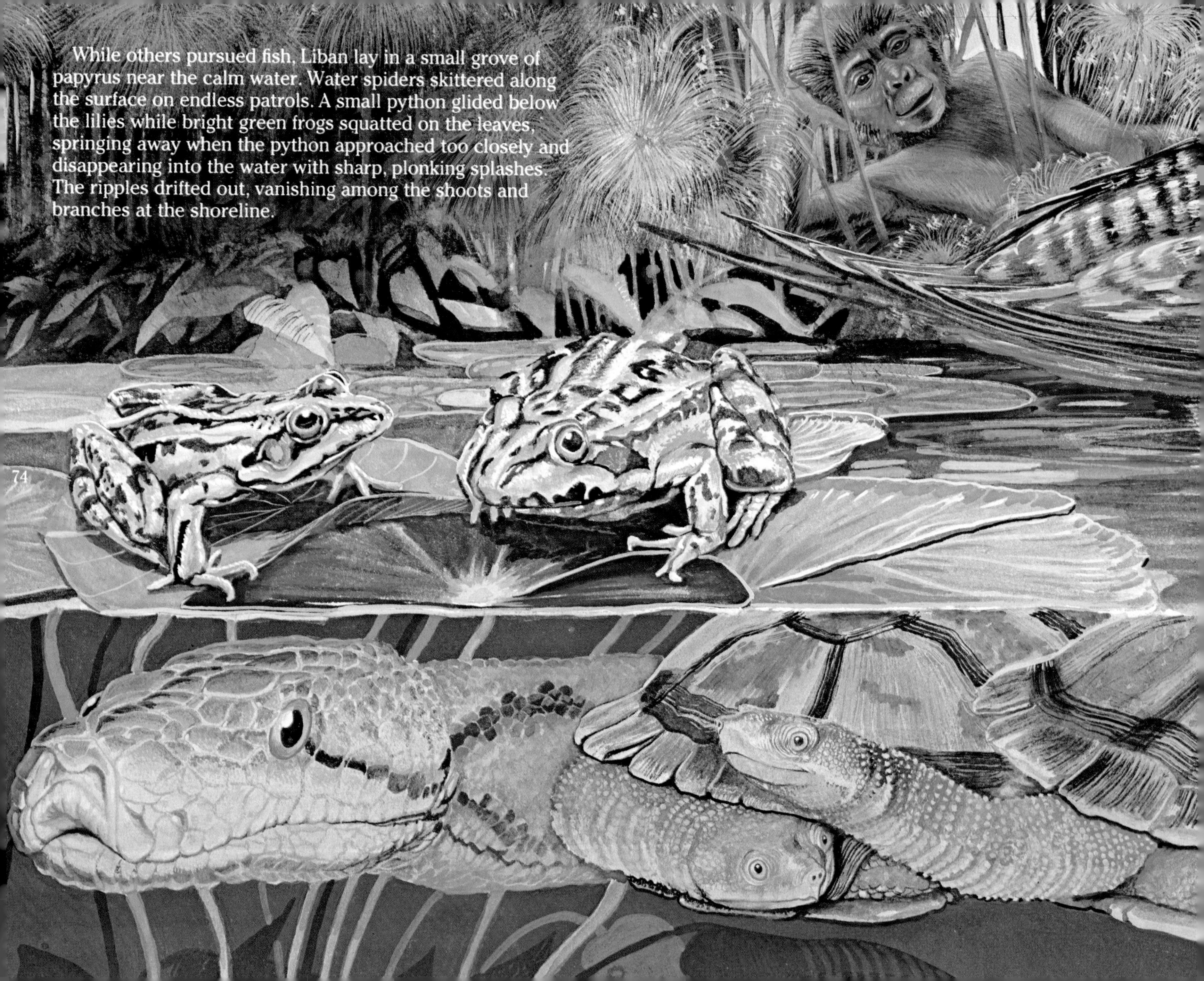

Far off in the trees a faint trilling could be heard coming closer. With a muffled purr of wings, two sandgrouses settled on the shore near Liban.

Ignoring the child lying still in the reeds, the female bird waded out along the edge of the water and turned to the male. He stepped quickly toward her, his head bobbing up and down. The female retreated, then turned, waiting for the male to approach again. He bobbed his head several times, cooed, and came on. She turned and waded a little farther, then turned back, watching him, and with barely a ripple was jerked out of sight beneath the water.

The female sandgrouse rose screeching to the surface but was immediately pulled back under. Liban peered down through the green water of the pool to see a number of turtles tearing at the bird. The water seemed to boil for several minutes. Then the last few feathers rose slowly to the surface of the pond.

The startling death of the sandgrouse sent Liban crying to her mother, who was contentedly nursing Lifi. Lucy pulled her close to the drowsing infant and began to groom her, massaging her and picking off parasites. Slowly, the little girl relaxed, secure in her mother's care. As Lucy continued to groom Liban, the other young children gathered around them and rested in the shade.

While the males were splashing around in the stream chasing fish, Lucy watched her mother direct the women and older children to search the isolated muddy pools back from the bank. The hot midday sun was quickly evaporating the remaining water, leaving many struggling catfish barely covered.

Using silt to get a firm grip on their slippery olive-green prey, the women and children brought the captured fish to

Eba and Cera, who stood by a large flat stone. While Cera held each fish firmly on the rock, Eba bashed their heads with a thick branch until they stopped moving.

As the others departed with all the fish that had been found in the pool, Eba grunted and pointed under the large boulders at the edges. Cera dug through the mud and, to her delight, pulled out three more clay-covered fish, which were quickly dispatched on the stone.

Baboons grooming

Monkeys and apes frequently pick through each other's fur to remove parasites and dirt. This behavior, called grooming, serves to keep animals clean; but, more importantly, it is a form of social communication. Such behavior reduces tension and promotes reassurance between animals of different rank. Cohesiveness is critical for survival of primates living in a social group and promotes transfer of learned behavior from one generation to the next.

The females and children returned to Lucy, their arms piled high with glistening fish. The males had returned with heaps of thick tubers, lily roots, nuts, turtle eggs, and Lonnog's large fish. The children scampered forward and placed their fish next to the food that the males had brought. Spread out at their feet was more than enough food for the entire group. The Family quickly gathered around the food to feast.

Lorcan knew that Lucy was particularly fond of the red roots from one particular plant. He placed five large ones in her hands. Lucy nodded with pleasure. She popped one into her mouth, pulling it forward to scrape off the tough outer skin before biting down into the crisp inner pulp. Lorcan squatted beside her and began, with much cracking and snapping, to work his way through a large pile of seeds and nuts.

In the meantime, Lonnog, taking his large fish down near the pool, sat with his back to the Family. This fish was his. He had caught it and was determined to enjoy it himself. He rubbed off the slimy coating, then, avoiding the spiny fins above and below, bit into its side. Slowly and with great relish, Lonnog ate every bit of the fish's flesh before tossing the skeleton into the pond.

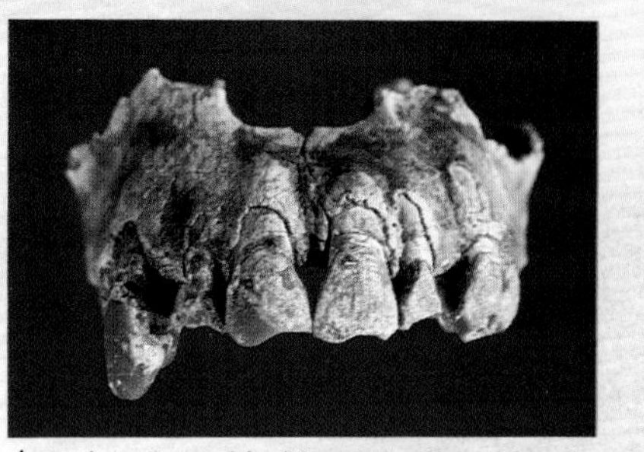

Anterior view of A. Afarensis *upper jaw*

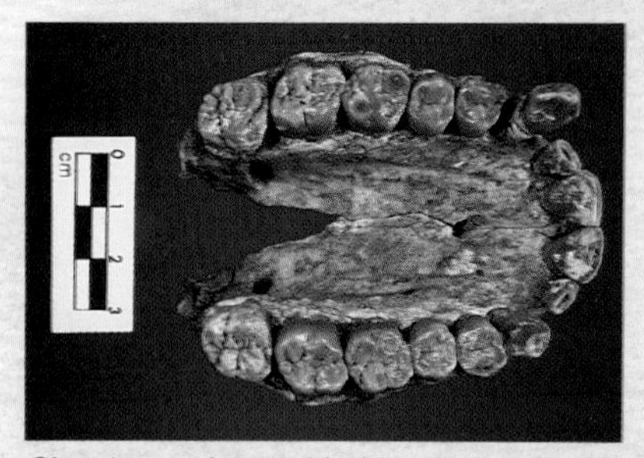

Chewing surfaces of A. Afarensis *teeth in upper jaw*

Teeth are the hardest tissue in the body and long outlast skin, muscle, and even bones after an animal's death. Teeth provide important clues to an animal's diet, and virtually every species of animal can be identified by its teeth.

Carnivores have large piercing canines and scissorlike back teeth, very effective for slicing and cutting up meat. Herbivores, like the horse, have large front teeth for cropping grass and broad, flat back teeth for crushing and grinding.

Hominids have a general-purpose dentition, reflecting an omnivorous diet. The front teeth—the incisors and canines—are for slicing and the back teeth for crushing and grinding. A scanning electron microscope was used to examine the chewing surface of Australopithecus afarensis *teeth. The majority of wear occurs on the front teeth. The right canine (slightly out of its socket) shows chipping damage from biting down on hard foods. Microstriations and the ribbonlike wear on the edge of the incisors suggests that they may also have been used to strip vegetation.*

After the meal, the Family rested on a series of stony outcroppings that projected over the river, ringed by large clumps of reeds. The highest spot of this group of rocks was casually occupied by Cano. After drinking from the river, Lorcan walked slowly up to where Cano was sitting above the rest of the group. The leader always occupied the highest position. To take it was a challenge to Lorcan's dominance of the Family. Lorcan paused a few steps away from Cano, then stretched and yawned while looking calmly out over the river. It was Lorcan's way of asking Cano to move without confronting him.

Lorcan knew that Cano resented him, that he felt compelled to challenge him. Cano felt he should have been made leader when Broc, his brother and the previous leader, had been killed by hyenas while defending Eba. But the Family, strongly influenced by Eba, had chosen Lorcan. As a result Cano remained jealous of Lorcan and resentful of Eba.

Cano gazed straight ahead, ignoring Lorcan's presence. Lucy and Cera looked at each other nervously. Since Eba had led the women to the large catch of fish, Cano sensed that Lorcan's position had been weakened. Alert for any weakness, Cano had chosen this moment to challenge Lorcan's primacy. Eba rose up and stood behind Lucy and Cera, tense and waiting.

Lorcan was not at all intimidated by the lounging Cano. He drew himself up to full height and stepped forward, interposing himself between Cano and the sun. He stared straight down at his rival. His lips grew tight over his teeth and his eyes flared. Cano continued to ignore him. Lorcan gave a low warning snarl. Cano looked slowly up at Lorcan and met the unflinching stare for a long moment. Then Cano faltered, his eyes glanced away, then back. His tight lips dissolved into a nervous grin. Lorcan did not move. After a few more moments, Cano uttered a low grunt. Lorcan took a half-step back. Cano rose, shoulders hunched, and clambered down.

Lorcan quickly sat in the vacated space and gazed out over the river. Cano made his way to where Dagan sat and evicted him in turn. The females below relaxed and went back to their ceaseless grooming of the children and each other.

Baboon yawning

This baboon is communicating a threat through nonverbal signals. The light-colored eyelids stand out dramatically from the surrounding dark brown facial hair and are clearly noticeable to other animals. In yawning the baboon exposes his large, and potentially dangerous, canine teeth. Animals recognizing such a threat will usually respond appropriately and avoid bodily contact. Large male baboons can combine threat displays, charges, and even severe bites to drive off animals that threaten a troop.

Except for the sulking Cano, the Family now rested contentedly. Lorcan made soft humming sounds from his perch. Lucy watched dense flocks of spoonbills and flamingos jostle and move about on the shores.

The flamingos especially interested her, with their dazzling color and their habit of walking backward, their long necks extended as their bills seined the bottom of the shallows. Lucy had never seen flamingos nesting, nor had she seen their young. To her, flamingos simply appeared—gleaming pink in their multitudes, descending without warning, flooding the river with their reflections.

On the other side of the rock, where the river ran deeper, Lonnog stretched out in a clump of reeds, gazing up at the sky. Beyond the reeds he heard a soft rustle and a splash. He parted the reeds and looked out.

An elephant with long curving tusks was plowing its way through the mats of vegetation that had built up in the backwater. On its back a white egret hopped from side to side. The bird's darting eyes searched the river weeds as the elephant plowed through them. From time to time, the bird would plunge off to spear a small frog or fish. The elephant wasn't at all disturbed by the egret. It benefited by the bird's constant search of its thick hide for leeches and insects.

Lonnog watched fascinated until a shift of the wind brought an acrid stench to his nostrils. Behind him he heard a series of soggy crunches. In a flash, Lonnog rolled to his right into a thicker patch of reeds, and a hippopotamus barged through the reeds, trampling the patch where Lonnog had been only seconds before.

The hippo was utterly unaware of the child it had almost crushed. Its small eyes in their protruding sockets were fixed firmly on the next bushel of reeds it was about to eat. A few slashes with its sharp incisors raked in a massive quantity of the thick papyrus as Lonnog, hidden only inches away, froze.

Lonnog hardly dared to breathe as an angry bellow sounded across the river. The hippo next to Lonnog opened its mouth to reply, exposing its thick, curving teeth. The next second the hippo trumpeted its response to the arriving male and marked the territory on which it stood as its own. A great spray of yellow liquid jetted from its hindquarters and was scattered by its small tail over Lonnog. The boy clutched the reeds tightly and remained motionless as the vile rain fell all over his body.

Bellowing, the hippo charged out of the reeds and into the river. Lonnog, reeking of hippo urine, jumped back deeper into the grass and shook himself violently, surrounded for a moment with a gleaming yellow haze. Then he peered out to watch the two beasts charge each other. Halfway through the first charge, both hippos stopped some yards apart. They threw back their heads and bellowed louder and louder at the sky. Then both opened their jaws and lunged, ripping and stabbing. Their bellows echoed from every tree. They backed away and charged again. This time their jaws locked together. Twisting, they rolled out of sight beneath the surface.

Lonnog had had enough of this battle. Taking the opportunity, he sprinted out of the reeds and leaped onto the rocks to regain the security of his Family.

Although Lonnog was happy to be close to the Family again, the feeling was not mutual. As the wind shifted they became aware of how strongly he had been marked by the hippo. They all backed away quickly, leaving the boy in the center of a very large circle.

Lonnog stopped, confused, and approached his mother. Lucy was having none of it. She backed away even farther and made it quite clear that Lonnog would have to bathe before he would be accepted back into the Family.

Most animals do not leave behind any trace of themselves when they die. Their flesh and bones are scattered and broken up, ultimately to disintegrate. If the dead animals are swept into a lake or a river and buried in sediments rich in certain chemicals, such as alkaline salts, they might become fossils. Through a chemical process, the organic material in the bones and teeth is slowly replaced, molecule by molecule, by inorganic elements such as calcium and silicon. This process, called fossilization, produces an exact stone replica of the original bone.

Scientist chipping sandstone from Hadar hippo cranium

Because their skin loses moisture at a very high rate, hippos spend most of their time in water, coming out at night to feed for five or six hours.

During Lucy's time, a smaller, different type of hippo lived in the lakes and streams. Instead of having only four incisors, it had six. Its face was shorter and its eye orbits did not project as much as in the modern hippo.

Many of the fossils, like this primitive elephant found at Hadar, were remarkably complete. This animal apparently had fallen into a lake and, before it disintegrated, was covered by sand and became fossilized. With such enormous, twisted tusks, this must have been an impressive animal to the early hominids.

Elephant excavation at Hadar

After Lonnog had washed the smell off, Lorcan led the Family back into the woods. Eba, still limping from her twisted ankle, trailed behind the others. This afternoon, although none noticed it at first, Cano was hanging back as well.

Unable to displace Lorcan, Cano turned his resentment toward Eba. She looked up to see him standing in her path, snarling. Cano's brow lowered over his eyes, taking them into deep shadow. With lips closed, his lower jaw began to clench tightly and his teeth began to grind. Eba stared back at him. He reached out to a thick sapling and snapped it off with a quick movement. Then, with a low but rising growl, he began to shake it closer and closer to her face. Eba did not move. Behind him, Lucy and Cera reappeared from the undergrowth, clutched each other, and began to make alarmed, soft hoots over and over.

The brush crashed and a shriek split the silence as a thick branch snapped behind Cano. He turned to see Lorcan, enraged, waving a stick high over his head. Grinning fearfully, Cano dodged around Eba and ran for the forest with Lorcan in rapid pursuit. He let out a high-pitched scream of surrender, but Lorcan bounded after him, the stick flailing.

Eba watched while Lucy and Cera screamed, and the children came tumbling out of the bush to see what was going on. Cano and Lorcan disappeared into the gloom of the underbrush. After a short while the brush parted and Lorcan walked calmly into the clearing.

The females and children clutched each other and whimpered. Far away, they could hear faint crashing sounds as Cano ran deeper and deeper into the grove. Lorcan tossed the stick aside and reassured the Family by draping his arms around them. Slowly, their whimpers calmed into soft panting and grunting sounds. Lorcan then walked on, and the Family followed. Cera was the last to leave the clearing. She looked back for a moment at the spot where her mate had disappeared, then turned and followed the others.

Dian Fossey with mountain gorilla

Until her tragic death in 1985, Dian Fossey dedicated nearly twenty years of her life to the study and preservation of the mountain gorilla. This photo was taken just minutes after Peanuts, a large male gorilla, became the first animal to reach out and touch Dian's hand. Gorillas beat their chests to announce their presence, signal other animals in the troop, or express intense excitement. Dian's work helped to preserve this threatened species.

Leaving Cano behind, the Family proceeded from the island toward the grove on the far side of the delta. Brig assumed her brother's place at the outside of the tightly grouped Family as they waded through the marshes.

After emerging from the muddy water choked with reeds and other vegetation, Eba noticed that the pain in her ankle was gone. She looked down and noticed that although the swelling was still there, the whole area felt numb. Then she saw a glistening black leech firmly engorged. Disgusted, she gently pulled the bloodsucker off and tossed it back into the marsh with a warning to the Family. They gathered together under a grove of acacia trees, grooming each other's fur in search of the wormlike parasites.

African spotted-backed weaver nest-building

It is easy to see why these birds are called weaverbirds. Representing a diverse group of birds found throughout Africa and Asia, each species weaves its own characteristic nest. Above, an African spotted-backed weaver is building a globular nest.

As the others picked and probed for leeches, Lucy's attention was drawn to a peculiar little nest that had fallen from a tree above, filled with scores of colorful, noisy weaverbirds. Pushing her finger into the long tubular opening, she found that the pouch was empty at its base. She picked up a few small pebbles, dropped them into the tightly woven structure, and watched them rattle together inside. Then she poured them out and repeated the process. This little distraction ended with a hoot from her mate, who was already leading the others into the grove.

The Baya weaver of Nepal builds peculiar elongated nests. It is a tantalizing notion to think that early hominids—watching these birds—may have gotten the idea to make baskets.

Baya weaverbird nest in Nepal

Walking into the grove beyond the delta, the Family was startled by a strong tremor in the earth, then another jolt, followed by cracking and splitting sounds. For a moment they huddled together fearfully. The vibrations stopped as suddenly as they had begun. Lorcan stretched himself up and sniffed the air, then made a quiet hoot. In the silence they could hear faint crackling and munching sounds. Lorcan moved off and the others silently followed him.

They came to the edge of a clearing nearby and saw a dinothere holding a thin sausage tree tightly between its tusks and trunk, shaking it violently. Seed pods drifted down in torrents. Releasing its grip on the tree, the animal turned and sucked up the pods with its trunk, bringing them to its mouth. It turned back to the tree and shook it so heavily that small fissures opened in the earth around its base. More pods rained down. The animal then used its tusk to strip a large hunk of bark off the tree. Its trunk wrapped around this strip, and it began to stuff the bark into its mouth. While the animal was eating the bark, Anu, Aine's fourteen-year-old daughter, scurried behind it unnoticed and gathered up an armful of pods. She repeated this maneuver twice before the beast noticed her and bellowed a challenge. Munching the seed pods, the Family faded back into the undergrowth as silently as they had arrived until they reached a clearing where they could safely rest.

As the Family lolled in the grove, Lorcan heard a faint humming. Intrigued, he stood up and gestured to Lonnog. As they walked through the sparse brush the humming became louder. Finally, they spotted a thick, decayed stump. The stump seemed to be singing. It was one of the rarest treasures of the forest: a beehive close to the ground. Lonnog began to move forward to raid the hive, but was stopped by his father. It was enough to know, for now, that the hive was there. Perhaps later, in the evening, when the bees were resting, they would return to take what they could of the honey.

The function of the five-foot-long, down-turned tusks of the Dinotherium *has always puzzled scientists. Unlike elephant tusks that sweep upward from the upper jaw and skull, these tusks protrude downward from the lower jaw. Scientists have found some clues in the dinothere's teeth and the relation of the creature's skull to its neck. Unlike an elephant's teeth that are used primarily for grinding and crushing, the forward teeth in a dinothere skull are used for crushing while those behind are for shearing. The skull had greater mobility than that of an elephant. The movement of the tusks were further enhanced by the pivoting action of the lower jaw. These clues hint that the creature probably used its tusks for more precise, delicate, and diverse functions than its bull-dozing contemporary counterpart. The dinothere's tusks probably served a variety of purposes including male rivalry battles, browsing and rooting functions, and, quite possibly, for shaking trees and stripping bark.*

Dinotherium *tusks*

Lorcan and Lonnog rejoined the others and they set off again. Soon the Family emerged from the grove and found itself on the savanna, where golden swaths of waist-high grass rolled off in all directions to the far horizon. The landscape teemed with life. Swift birds darted around the Family as their movement caused grasshoppers and other flying insects to whirr up out of the grass.

Near a clump of tangled brush, Liban paused to study the skeleton of a long-dead antelope. From the curved horns of the beast hung a host of strange, translucent cocoons. One was not translucent, but dark. As she watched, it moved. A small moth, its wings folded tight against its body, struggled out. It crawled along the underside of the horn and rested at the tip, letting the sun draw the dampness from its wings as they unfolded.

Liban tilted her head to one side and the other. The moth beat its wings once, twice, then dropped down and fluttered back up. Liban shot her hand out to grab it, but the moth dodged and flew higher. She hooted in delight as it fluttered down onto a flower in a bush a few yards away.

She joyfully danced after the brilliant creature that perched on the flower, but as she approached it, the moth simply vanished.

When Liban got to the flower she looked carefully around, and then into a single dark eye peering back at her. Attached to the eye was a bright green chameleon with just the tip of a moth wing protruding from the corner of its mouth. As Liban watched, a bright dragonfly hummed past her head. Suddenly a jet of pale yellow shot from the chameleon's mouth, stopping the dragonfly in mid-flight and reeling it back into the chameleon's jaws. The dry crunch of the chameleon's meal dispelled the magic of the moment. Turning away, she ran after the foraging Family.

Chameleons have a set of specializations which permit them to trap elusive flying insects. Their opposable feet allow them to crawl securely on branches. Their curled, prehensile tail is also a benefit for gripping. They sway slightly as they move, mimicking a moving leaf, and their sideways-compressed body even makes them look like a leaf.

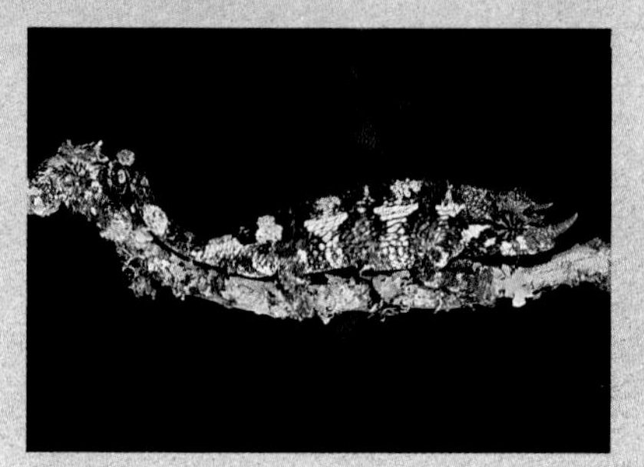

Chameleon mimicking lichens

Chameleon mimicking green background

Synonymous with chameleons is their ability to change color to mimic the immediate environment. The pigment cells in their skin even change color if the chameleon is blindfolded. In one color phase, this Jackson's chameleon is mimicking the lichen-covered branch; in another, it matches the green leaves.

A great asset in hunting is the ability of each eye to move independently, searching for prey. Once a likely subject is found, the other eye rotates and, with binocular vision, fixes the prey

Chameleon with baby

in space. The jaws open and in a twenty-fifth of a second the tongue, which is stored in the mouth like a concertina, shoots forward. The tongue may be as long as the body and tail com-

Chameleon catching insect

bined, and has an adhesive tip with mucous glands. With lightning speed the tongue snaps back, carrying the meal into the mouth.

Some African moths have the special habit of laying their eggs in the cracks and crevices of antelope horns. The larvae burrow into the horns and feed on them. When the insects pupate they leave their empty cases projecting, as seen here, on the horns of an eland skull.

Antelope horns with insect cases

Lonnog had the sharpest eyes of the Family. He looked slowly and carefully around the land. Here and there herds of wildebeest and gazelle grazed. Packs of big cats and wild dogs rested in the shadows of distant rocks. A tightening circle of black vultures marked a kill. In the distance, three pillars of mud towered over the grass.

Lonnog pointed toward the mud towers and grunted with pleasure. To him and the Family these pillars meant food. They were termite mounds.

With a whoop, Lonnog and his cousins Bran and Ciar started off toward the mounds at a steady lope. As the youths sped across the savanna, Bran stopped and called them back with three sharp hoots. He'd almost run straight into a strange animal. It was a pangolin busy digging up a nest of honey ants. The girl and two boys returned and clustered around the heavily armored creature. In fear, the pangolin rolled itself into a tight, scale-covered ball.

Bran edged up and poked it. It rolled away on the hard clay. He closed in and gave it a harder kick toward Lonnog, who booted it back to Bran. Seven-year-old Ciar was not impressed and left the boys with their sport. Behind them, the rest of the Family had come upon the ant nest that the pangolin had ripped open and were helping themselves to the honey ants.

Pangolins are called scaly anteaters because the top side of their bodies is covered with brown, tile-like scales, giving them the appearance of a pine cone. These leathery scales can be closed tightly when the animal rolls into a ball, as one is doing here to defend itself from the curious lion cub.

Its claws are used to break into termite mounds and ant nests. Its elongated tongue, used to lick up ants and termites, is kept moist by large salivary glands. Since their diet requires no chewing, pangolins do not have teeth.

Pangolin in a ball with lion cub

Since the pangolin had stripped away the surface, it was easy for the Family to get to the tasty worker ants. These ants had large swollen abdomens, which contained a sweet fluid milked from great clusters of aphids they cultivated on the grass and brush above the nest.

Lorcan, Dagan, the females, and the younger children squatted around the nest with much smacking of the lips. Lucy pulled Liban close and had her watch. She picked up a honey ant, carefully avoiding the sharp mandibles, and nipped off the amber abdomen filled with sweet fluid. The honeydew seemed to replenish a lot of the energy that the mother had lost last night in childbirth. Eba, Cera, Lorcan, and Dagan squatted on the far side of the nest eating their way through the bustling worker ants with great satisfaction.

Liban plucked up an ant but dropped it in shock as its pincers nipped her finger. She quickly lost interest in this treat and began to examine the brush above. Here the workers were going about their business. An ant would march up to an aphid, stroke it gently with its antennae, and wait. The aphid would then emit a single gleaming drop of honydew, which the ant would drink. The process was repeated until the ant's abdomen had swollen with the fluid. Then the ant would return to the nest below. Of course, on this day few of the workers would arrive at the nest.

Honey ants

These honey ants live in dry country, where food and water may not be available for long periods. As a precaution against such lean times, these ants store liquid food called honeydew in worker ants, which is regurgitated when needed.

Ants tending aphids

Ants are remarkably social insects. It has even been documented that some kinds will tend other insects. The ant hosts protect aphids from predators and parasites in exchange for honeydew produced by the aphids. Feeding on plant sap, the aphids pass through their body a liquid called honeydew, very rich in sugar. Ants will stroke the aphids with their antennae, stimulating the aphid to excrete the honeydew.

After tiring of their game with the hapless pangolin, the boys headed directly to the termite mounds. Young trees and bushes sprouted from the first mound. The next mound was heavily eroded and riddled with holes. The termites had abandoned it, but it was a new home for numerous other creatures. However, the center mound was well over ten feet tall and solid, indicating that a large termite colony still dwelled inside.

A snarl from behind the center mound made the youths turn and scamper back to the adult males, Dagan and Lorcan, who had just arrived. Peering carefully around the mound, the males saw a small, fierce ratel rooting at the base for the small mammals, reptiles, and insects that now inhabited the abandoned chambers.

Although it was small, it was fearless, and the Family had a good deal of respect for ratels. They watched as the wolverine-like creature flushed a scorpion, avoiding the deadly tail stinger, and, with a quickness that startled them, killed and ate it. Only the stinger was left, twitching in the dirt. The ratel turned back to the mound in search of another morsel and saw the males of the Family. Snarling viciously, it jumped forward, its hackles up and its sharp teeth fully displayed. The males responded by swinging their sticks at it with a chorus of growls. Intimidated by their numbers, the ratel retreated, snarling. The males advanced, this time putting the pugnacious animal to flight. Lonnog, emboldened by this retreat, moved after it. A few yards away, the ratel turned once more and bared its teeth. Lonnog, thinking better of his plan, returned to the safety of the adults.

Termite mounds in Sudan

Termite mounds are familiar structures on the African landscape. This tall, spirelike mound from the Sudan is active with millions of termites.

When termite mounds are abandoned, like this one at Lake Manyara in Tanzania, rain erodes them and they may become anchors for growing trees.

Eroded termite mound

Soon the Family was gathered around the inhabited termite mound. Bending down to the foot of the mound, Lorcan began to sniff around it. Lonnog, imitating his father, did the same.

On the far side, the dusty smell of earth gave way to a sharper, more acrid odor. Leaning closer to the base, Lonnog saw that the normal patterns of erosion around the mound gave way in this one instance to a pathway smoothed out by hundreds of thousands of tiny feet. He followed this path very carefully on his hands and knees until he saw an opening at the base of the mound.

Lonnog hooted twice. Lorcan was by his side in an instant. Knowing immediately that they had found the main entrance to the mound and that there was a rich lode of termites within, Lorcan tore off a slim blade of grass. He moistened this in his mouth and then, with great delicacy, slowly slid it into the aperture. After a moment, he withdrew it. On the tip were four wriggling soldier termites, their pincers still locked around the grass. Lorcan plucked two off the tip and popped them into his mouth. Smacking his lips with great relish, he offered the remaining two to Lonnog. Then Lorcan moved to one side and began to knock out hunks of earth with his stick. He carved his way in toward the center of the nest. Dagan, Lucy, Eba, and Cera soon joined him. In almost no time hunks of the mound piled up at their feet.

Hundreds of worker termites carrying small clods of moist clay rushed up from the center of the nest to repair the damage. Lorcan waited until the opening was choked with white bodies, then scooped them out onto the ground to be shared by all. Again he waited for the hole to be filled, and scooped again until the earth was crawling with the insects.

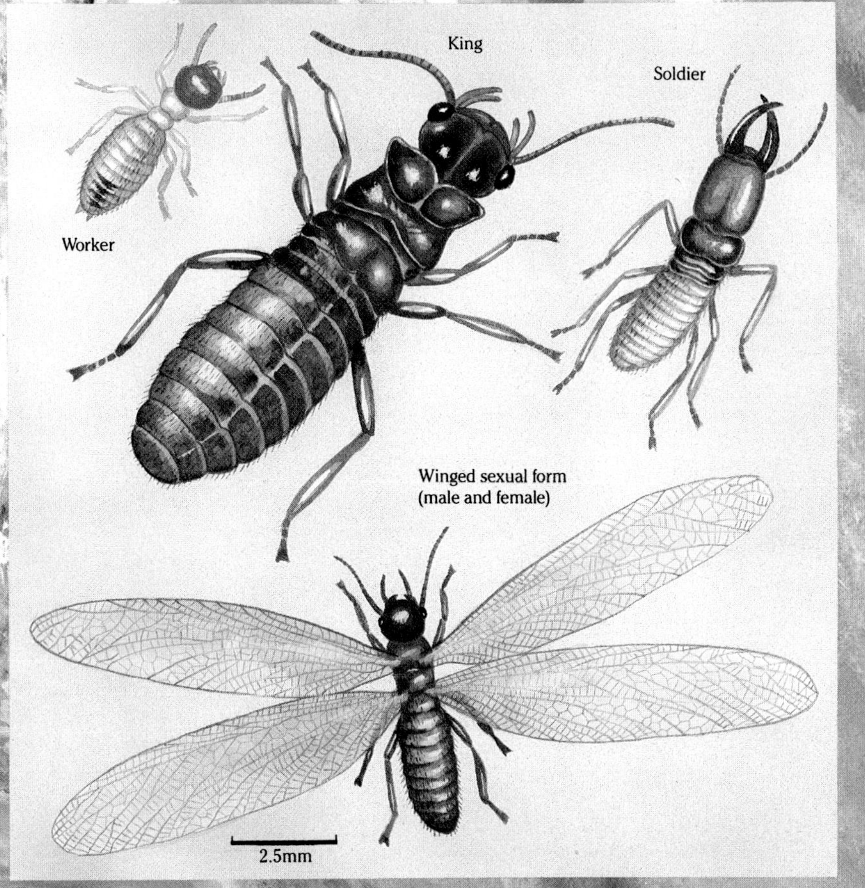

Termite types

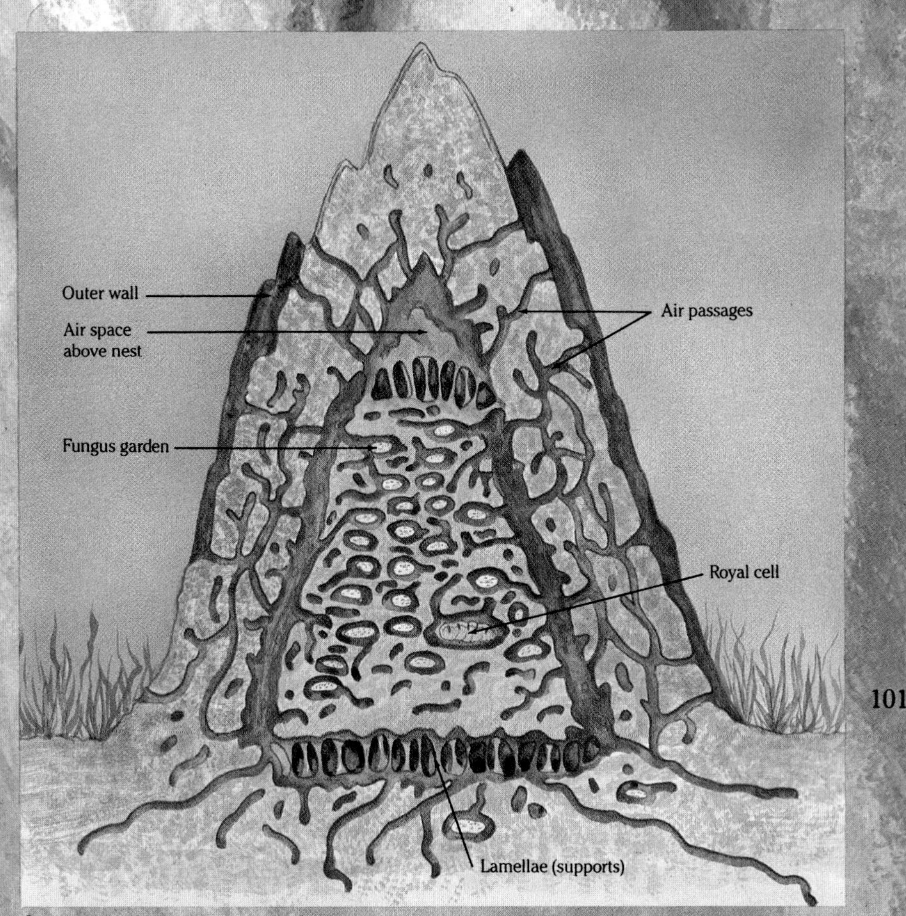

Cross section of termite mound

The hot air in the central core of the termitary, where the major portion of the colony lives, rises upward. It migrates into chambers near the outer wall of the mound, where it is cooled. Then the cooled air sinks back to the center to begin recycling again, providing natural air-conditioning.

Some termites can digest wood; other species actually cultivate their own food. Fungus that grows inside the mound is eaten by the termites. The fungus is nourished by partially digested wood pulp regurgitated by the termite.

Termites are organized in a caste system: foragers bring food into the mound; soldiers protect the mound; builders construct the mound. A queen termite, reaching four inches in length, may produce 30,000 eggs per day for up to twenty years. Workers carry the eggs to chambers where they will hatch into nymphs.

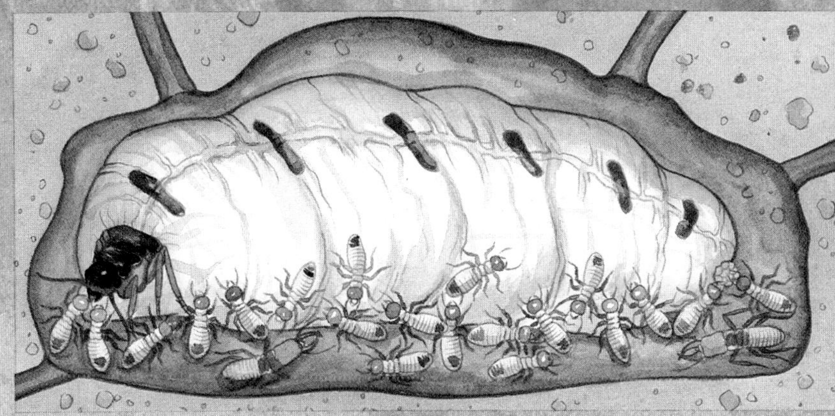

Queen's chamber

Lucy and Lorcan lazed against the termite mound feeling the warmth of the sun bake their well-fed bodies. Liban came over and climbed into her father's lap as if to mimic the infant that usually nuzzled Lucy's breast. Lorcan reached down and absently began to groom his daughter with one hand. With the other, he plucked another blade of grass and began to probe the mound. After a moment he withdrew it and let it dangle in front of Liban. The little girl looked at the termites that clung to it. Not hungry, but not one to pass up a treat, she chewed them slowly. Lorcan pointedly handed the blade to Liban. Liban got out of his lap and gave the blade to Lucy, who handed it back to her daughter. Liban looked at the blade and then at the opening in the heart of the nest. She hesitated for a moment, carefully pushed the strip of grass deep into the hole, then pulled it out. This time there were six termites dangling from it. Liban gave a squeal of delight and waved the blade around in a triumphant gesture before eating her first catch one by one.

Chimps termiting

For centuries humans have been singled out as unique because of our ability to make and use tools. However, in 1960 Jane Goodall, working at Gombe National Park in Tanzania, observed chimpanzees carrying grass stems and sticks, sometimes over long distances, to a termite mound. They scrape open a passage in the mound and slowly insert a grass stem or twig. After a short pause, they carefully withdraw the grass stem. Using their lips and teeth, they pick off the termites and eat them.

In other parts of Africa, chimps have been observed using stones to crack open different kinds of nuts. Chimps have also been seen using chewed up leaves as a sponge to obtain water from a hollow in a tree.

The use of tools by chimps suggests that Lucy and her kind may have used simple, perishable tools, which would not have been preserved in the geological record.

Bran, Lonnog, and Ciar, followed by her younger brother, Ocan, scampered up the hillside beyond the termite mound, leaving the others below. Above them a white plume of steam accompanied by a hissing sound attracted the young group. When they reached the crest, the earth below them opened into a steep ravine. Steaming water cascaded down the rocks, filling the gorge with a gray mist. Bran peered down through the foggy clouds and saw a large pool at its base surrounded by rainbow colors. Intrigued, he jumped down the side of the ravine and onto the slick boulders beside pools of scalding water. Lonnog, following the older boy, helped Ciar lift her little brother down the rocks.

The steep walls and boulders of the ravine were encrusted with thick calcium deposits. Dark green mats of algae, slippery from the spray, covered the boulders on all sides. Flies swarmed on the mats, rising as the children slipped past them. After a number of falls, the children held tightly together. They squinted through the thick black clouds of flies that grew denser as they got closer to the bottom. During their descent each pool became increasingly cooler, until they reached the large pond at the base which was pleasantly warm.

Slick gray boulders dotted the pond. Lonnog, trying to impress his older cousin, leapt out onto one in the shallow water. The rock shuddered and sank out of sight, tumbling the boy into the warm water. The rock rose back to the surface, and the unconcerned eyes of a large tortoise blinked calmly at the soaked boy. Wading back to the shore, he and the other children watched scores of tortoises resting and drinking in the pool before setting out on their isolated treks across the grasslands. Bran knew that the reptiles were harmless, and he relaxed in the water while Ocan and Ciar rode about on the backs of the tortoises.

In areas of volcanic activity, like the Great Rift Valley, underground rock often remains very hot for long periods of time. Water collected in cracks and caverns is heated, resulting in hot springs or geysers. Geysers occur when the subterranean water begins to boil and the pressure causes it to shoot into the air.

These thermal waters have high concentrations of sulfur, phosphates, and carbonates, making the water undrinkable and, often, poisonous.

Geyser with related calcium formations

Hot springs fumarole, or steam vent

A shift in the wind carried a stench like rotting eggs to the children playing in the warm water. The fun and lightness seemed to go out of them, but their curiosity drove them up and over the crest of a small hill.

As they emerged they noticed that the grass beneath them had turned dry and brittle and was disintegrating into a fine powder when their feet touched it. Before them was a small, shallow valley with a pool of water, pale yellow under the declining sun. The vegetation began as a pale green, but quickly turned yellow, then brown, and finally a chalky gray as it neared the pond. The shoreline was a yellowish, chalky substance with small rivulets of a milky orange liquid oozing up and leaking into the pond. Around the pond's edge were scores of small dead animals, their bodies drying in the sun. No flies or insects buzzed about on the carcasses.

Walking cautiously onto the strange landscape, the children became aware of the oddest thing of all—except for the distant hiss of steam, it was utterly silent. Nothing in their experience was as foreign to them as absolute silence. Even in the midst of the darkest nights on the savanna, animals moved in the bush, herds called out on the plains, insects chattered their secret songs, and the wind stirred the dry trees. Always, there was some sound, some pulse of the plains that let them know it was teeming with life. Here there was only the crunch of dead grass underfoot.

Suddenly a bird called out above them. They looked up to

see a large marabou stork sweep down to feast on the bodies of the animals by the lake. It flew over them, circled to the right along the shoreline, and splashed down into the pool with its beak ready to snap up a marmot's swollen carcass.

As soon as the stork's legs entered the water, it gave a shriek and began to beat its wings in a frantic effort to reverse itself. The wings splashed into the water and more of the burning liquid settled on the bird's feathers and its eyes. The bird began to shriek desperately and frantically beat its wings. But it was too late. In a few moments, as the children watched in horror, the stork toppled over and began to thrash about in the pool, slowly dying as the water ate at its flesh. It made several attempts to right itself and stagger out of the water, but each time it fell back. Finally it did not try to rise again but simply lay in the pool, twitching and calling out until it fell silent and did not move again.

The children, mesmerized by this spectacle, became aware that their own feet seemed hotter, almost burning. With a series of panicked hoots and cries they ran from that terrible place as fast as they could move. With every step, the burning of their feet grew worse and worse.

They reached the crest of the hill and with cries of pain flung themselves down the steep slope toward the tortoise pond. The children plunged into the pond, where the warm water washed away the terrible acid burns of the poisoned ground.

The element sulfur can become concentrated by the thermal activity associated with volcanoes. The sulfur combines with water to produce sulfuric acid. In high enough concentration, a freshwater lake turns into an acid lake, killing any living creature that ventures into it.

This view of Grand Prismatic Spring at Yellowstone National Park shows dramatic colors of mineral deposits and living algae which may resemble some of the earliest life forms on earth.

Hot sulfur pool at Yellowstone Park

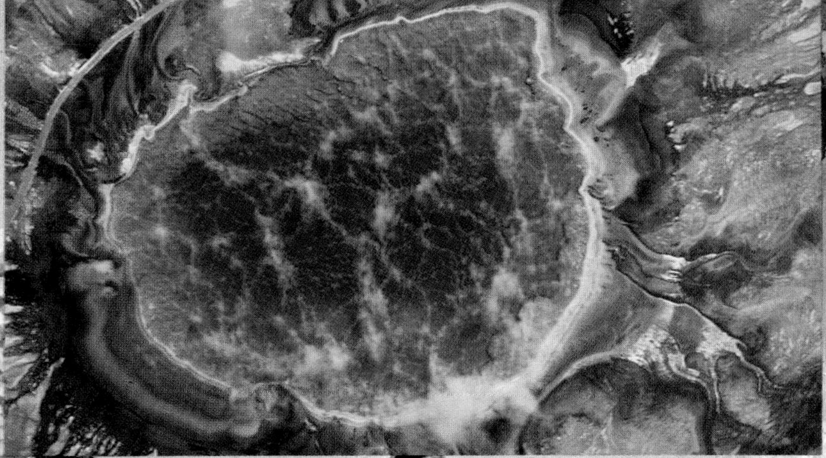

The four youths made their way back toward the savanna. When they reached the flatlands, a large cloud of dust grew on the horizon. Glancing over his shoulder at this cloud, Bran suddenly broke out of his loping walk and ran quickly toward a pair of isolated acacia trees. Ciar and the two boys followed at a run. Behind them, the cloud began to resolve itself into a horde of dots that became hundreds of small, zebra-like horses. The herd was on its daily migration to the safer parts of the plain where the grass was shorter and could not easily conceal predators.

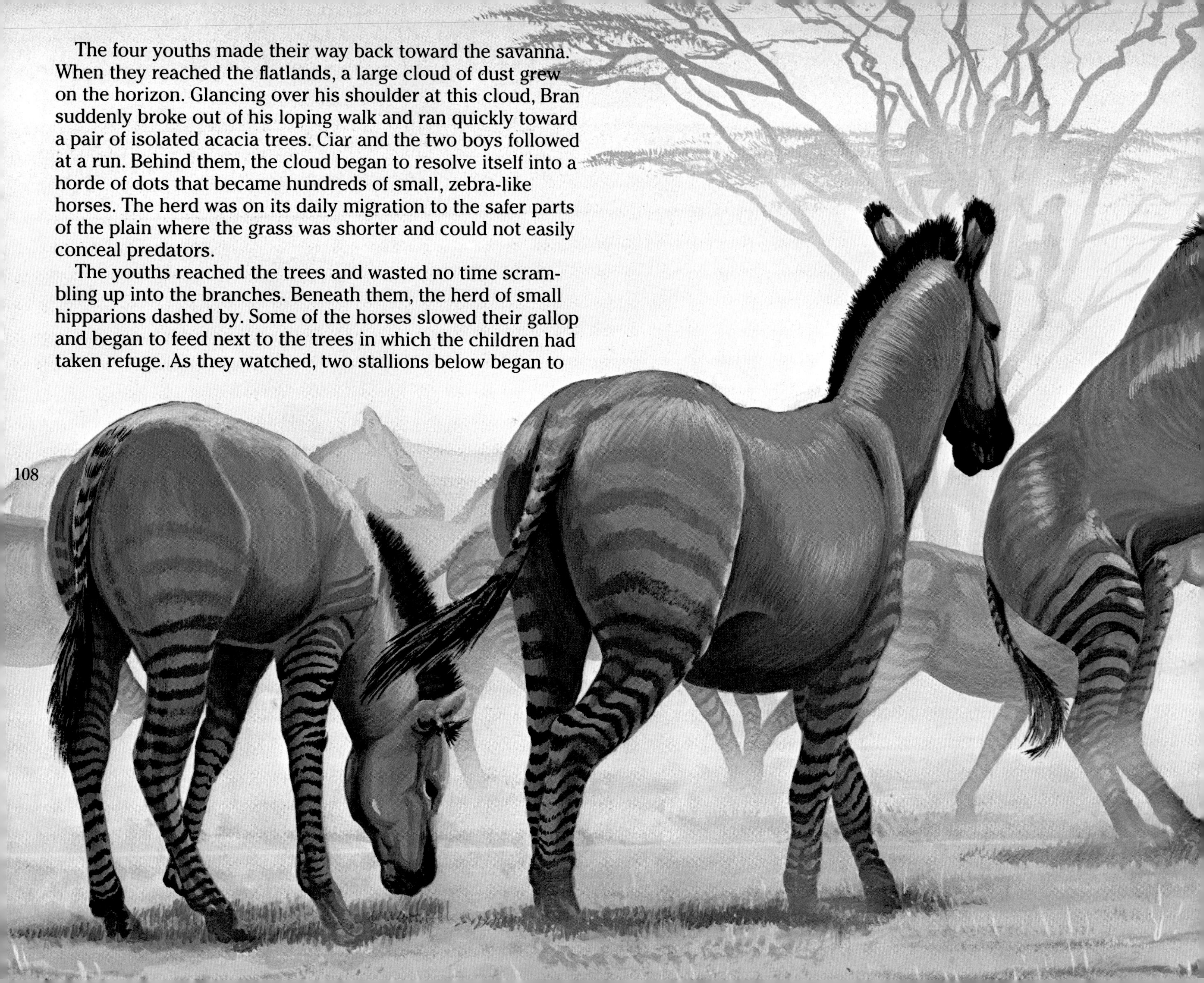

The youths reached the trees and wasted no time scrambling up into the branches. Beneath them, the herd of small hipparions dashed by. Some of the horses slowed their gallop and began to feed next to the trees in which the children had taken refuge. As they watched, two stallions below began to fight over a nearby mare. One took a vicious nip at the flank of the other. The other struck straight out with his hind legs and landed a solid blow in his opponent's side. As they fought, they stumbled into the mare. Furious, she turned toward the fighting males and drove them both off. A few minutes later she trotted after the herd out onto the broad plains of the savanna. After the horses were gone, the children climbed down from the trees and set off again toward the grove.

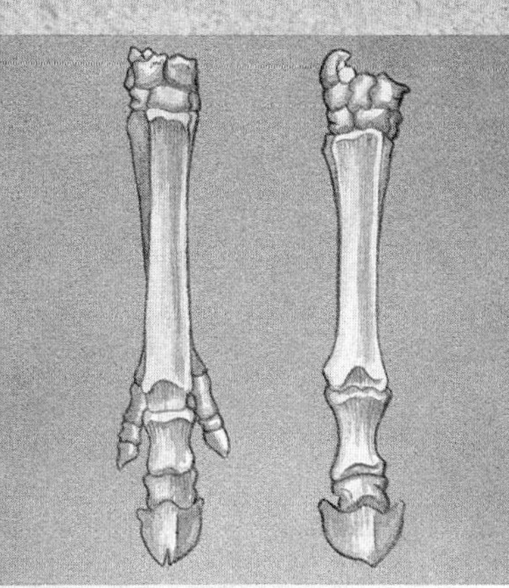

Hipparion *foot on left, compared to a modern horse foot*

Animals with hooves are divided into odd-toed forms—horses and rhinos; or even-toed forms—giraffes, antelope, pigs, camels, bison, and cows.

In Lucy's time, horses, or Hipparion, *were much smaller than those of today. They had three toes, instead of the single toe (hoof) of the present-day horse and zebra, which are called* Equus. Hipparion *was not the ancestor to modern horses, which evolved in North America from a totally different form.* Equus *migrated into Asia and Europe, finally reaching East Africa about 2.3 million years ago.*

The Family had already begun to establish camp when the youths arrived. The females had folded down large clumps of grass and brush and arranged them into sleeping nests. The females and children would sleep in clusters within the circle while the males rested at its circumference. Dusty youths, famished from their ordeals, made their way deeper into the grove for a snack. Within minutes, they came to a place where the brush and vines grew thickly. A rustling sound grew louder as they approached.

Ciar stepped carefully into the brush, poked her head cautiously around a thick tree trunk, and peered into the clearing. Close by, stretching up into the lower branches of a sausage tree, was a grazing sivatherium. Ciar and Lonnog stared at the tall giraffe-like creature that was methodically stripping a sausage tree of its seed pods.

Most primates have prehensile (capable of grasping) hands and feet useful in climbing.

Apes have a very shortened thumb and long fingers. They grasp things, much like a human baby, with a power grip, holding things between the fingers and the palm. Lucy's hands and those of later hominids are even more versatile. The fingers are shorter and the thumb longer, and able to rotate to reach all the fingertips. This permits use of the precision grip, the ability to hold objects between the tip of the thumb and the fingertips.

The striking difference between the human foot and the feet of apes and monkeys is that our big toe is not divergent. An ape foot is more similar to an ape hand than is the case in humans.

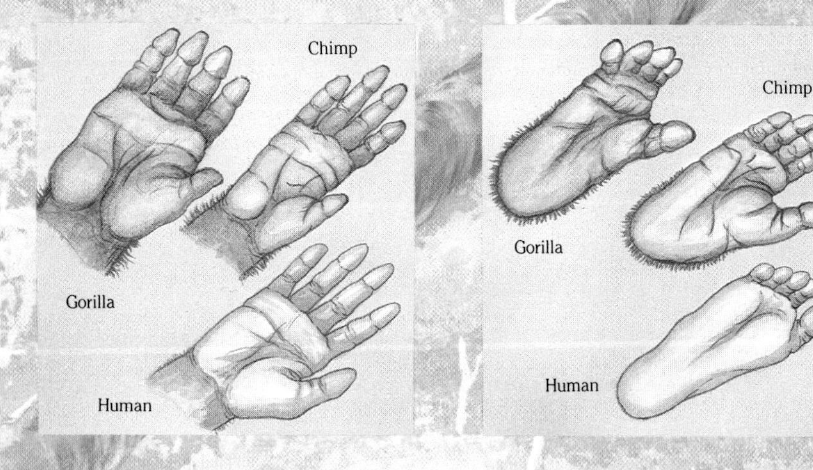

Although lacking the elongated neck of present-day giraffes, the sivatherium was still the tallest animal in the valley. Its ornate horned crest topped a body that was half okapi and half giraffe; its powerful haunches were thickly muscled. Its neck stretched up, giving the sivatherium the reach to pluck and slurp down pod after pod of rich seeds from the sausage tree. It ignored the small screaming monkey in the tree who was outraged at this intrusion into what it considered its private food supply.

Taking his cue from the monkey in the tree, Lonnog attempted to shinny up a thick vine into the high limbs of the sausage tree. But Lonnog was not as well adapted to tree climbing as the monkey. His hands and arms tired before he was halfway up and, frustrated, he slid back to the ground.

Fortunately the sivatherium was a sloppy diner and knocked many of the tender green pods to the ground as it ate. The children were quick to take advantage of this. Carefully sneaking in behind the beast they gathered up their snacks.

Chimpanzees prefer sleeping in tree nests rather than on the ground. By bending and interweaving branches, they build a firm platform on which to sleep. In the rainy season, a day nest is built for shelter.

Because of their heavy weight, ground nests are more common for gorillas. The nests are built of vegetation and look like oval, leafy bathtubs. The mountain gorilla pictured here is in repose on some soft vegetation.

Gorilla reclining

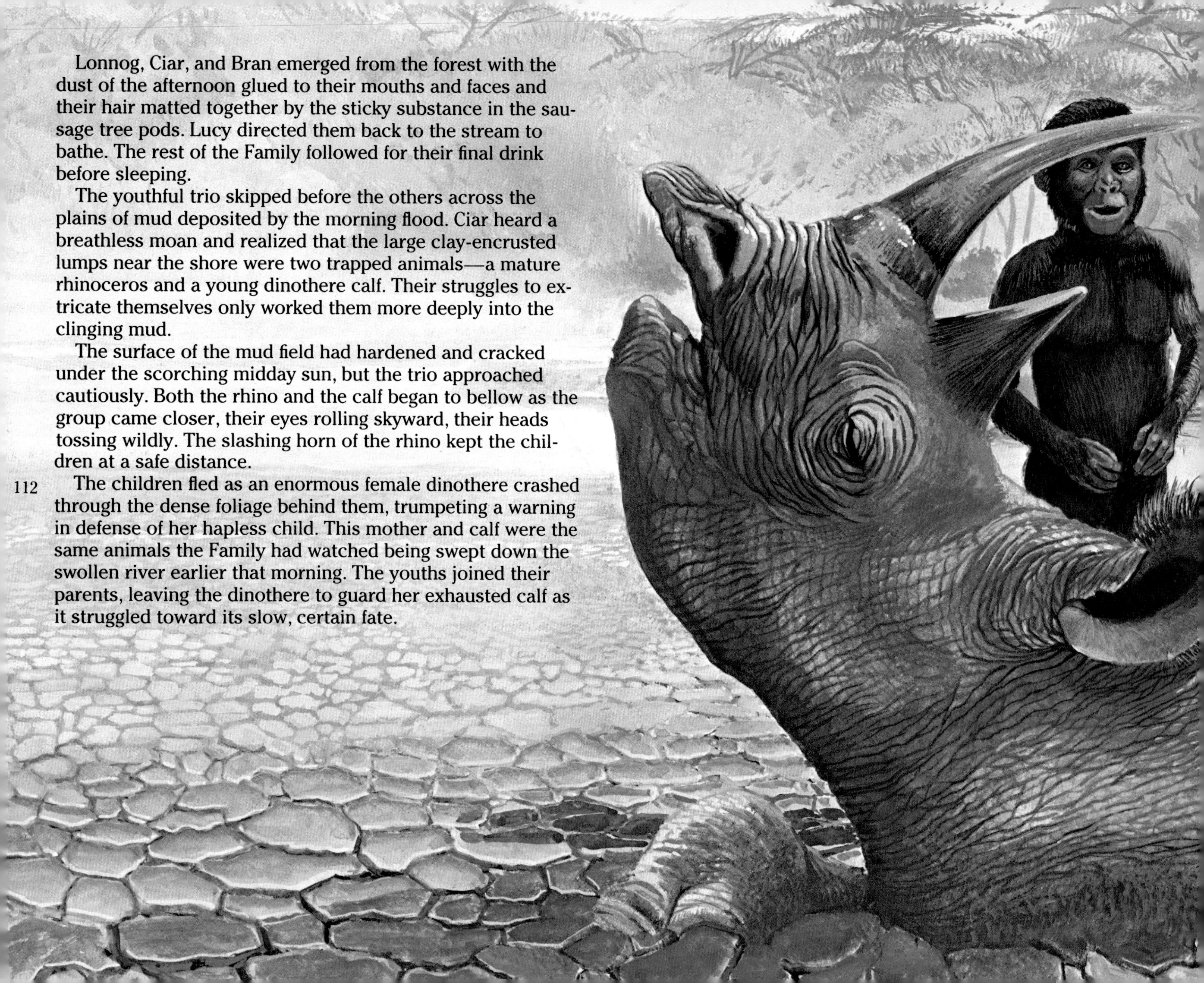

Lonnog, Ciar, and Bran emerged from the forest with the dust of the afternoon glued to their mouths and faces and their hair matted together by the sticky substance in the sausage tree pods. Lucy directed them back to the stream to bathe. The rest of the Family followed for their final drink before sleeping.

The youthful trio skipped before the others across the plains of mud deposited by the morning flood. Ciar heard a breathless moan and realized that the large clay-encrusted lumps near the shore were two trapped animals—a mature rhinoceros and a young dinothere calf. Their struggles to extricate themselves only worked them more deeply into the clinging mud.

The surface of the mud field had hardened and cracked under the scorching midday sun, but the trio approached cautiously. Both the rhino and the calf began to bellow as the group came closer, their eyes rolling skyward, their heads tossing wildly. The slashing horn of the rhino kept the children at a safe distance.

The children fled as an enormous female dinothere crashed through the dense foliage behind them, trumpeting a warning in defense of her hapless child. This mother and calf were the same animals the Family had watched being swept down the swollen river earlier that morning. The youths joined their parents, leaving the dinothere to guard her exhausted calf as it struggled toward its slow, certain fate.

With flooding, the soil may become so muddy that animals become trapped and are unable to move. The commotion caused by such an event is sure to attract the attention of carnivores and especially scavengers. The trapped animals become easy prey for carnivores and perhaps even hominids. While our earliest ancestors, like Lucy, were not well equipped—lacking stone tools to benefit from such an event—later hominids probably did exploit such tragedies.

The Family had stopped by a small brook leading into the river to bathe and drink. The brook ran broad and shallow under a thick canopy of ebony and sausage trees. In the shallow brook, small fish swam in thick schools. As the sun settled toward the horizon, Cera sat off to one side. In the hours that had passed since Lorcan had driven Cano off following his attack on Eba, Cera had become more disconsolate and lonely. Even now, she stood off to one side of the Family as if waiting for her mate to return.

Her expectation was realized when the foliage parted on

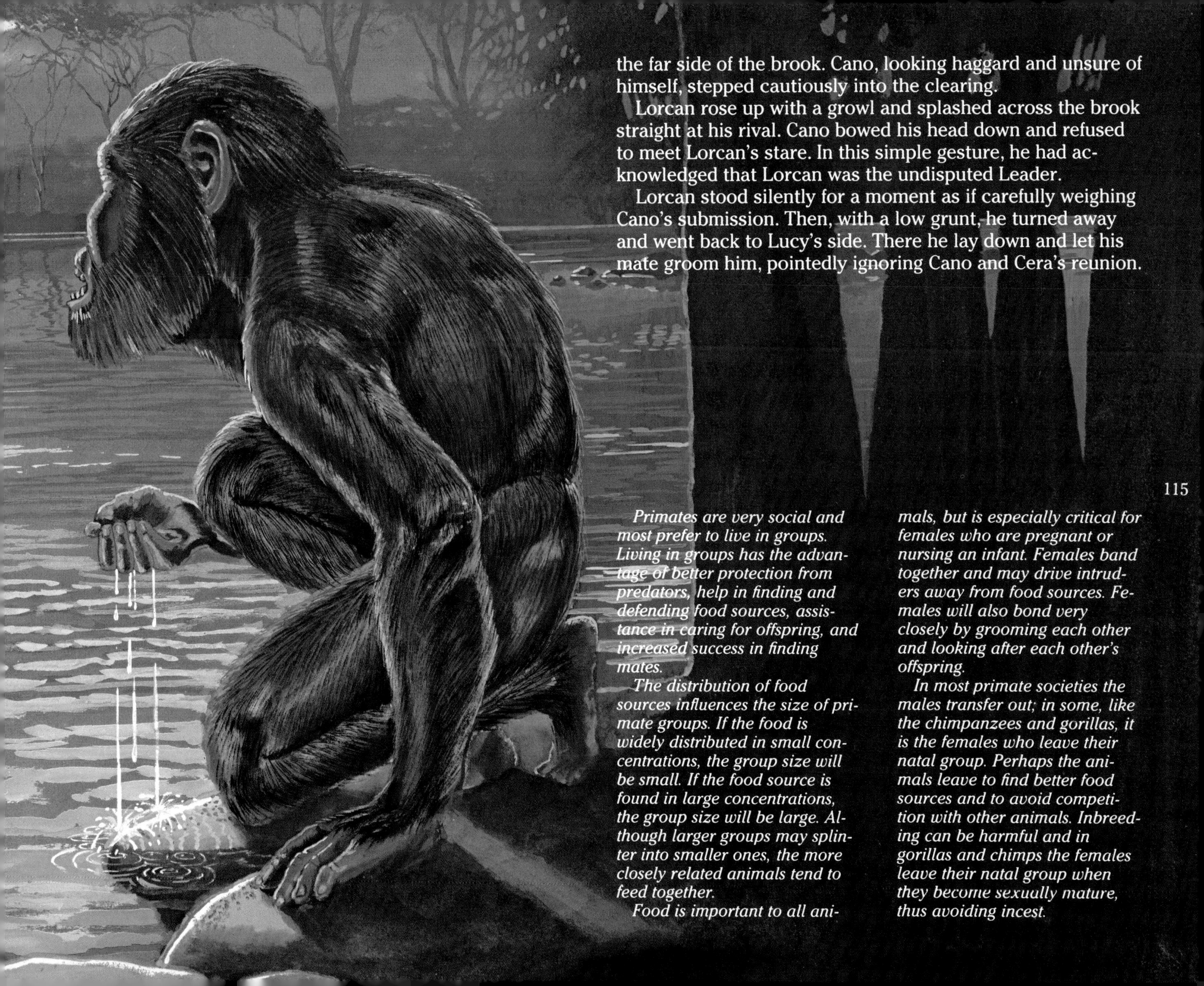

the far side of the brook. Cano, looking haggard and unsure of himself, stepped cautiously into the clearing.

Lorcan rose up with a growl and splashed across the brook straight at his rival. Cano bowed his head down and refused to meet Lorcan's stare. In this simple gesture, he had acknowledged that Lorcan was the undisputed Leader.

Lorcan stood silently for a moment as if carefully weighing Cano's submission. Then, with a low grunt, he turned away and went back to Lucy's side. There he lay down and let his mate groom him, pointedly ignoring Cano and Cera's reunion.

Primates are very social and most prefer to live in groups. Living in groups has the advantage of better protection from predators, help in finding and defending food sources, assistance in caring for offspring, and increased success in finding mates.

The distribution of food sources influences the size of primate groups. If the food is widely distributed in small concentrations, the group size will be small. If the food source is found in large concentrations, the group size will be large. Although larger groups may splinter into smaller ones, the more closely related animals tend to feed together.

Food is important to all animals, but is especially critical for females who are pregnant or nursing an infant. Females band together and may drive intruders away from food sources. Females will also bond very closely by grooming each other and looking after each other's offspring.

In most primate societies the males transfer out; in some, like the chimpanzees and gorillas, it is the females who leave their natal group. Perhaps the animals leave to find better food sources and to avoid competition with other animals. Inbreeding can be harmful and in gorillas and chimps the females leave their natal group when they become sexually mature, thus avoiding incest.

Liban continued across the golden fields. Sniffing the air she noticed a sharp, pungent odor, then saw a pulsing black line extending back as far as she could see. There was a piercing pain in her foot, then another and another and another. Horrified, she looked down and saw a mass of soldier ants crawling up her leg. She howled in panic as more and more ants swarmed onto her flesh. Miraculously, two hands lifted her into the air. It was her brother Lonnog, whose feet and lower legs were now thick with ants.

Stumbling from the bites he had received, Lonnog carried her to Cera's arms. Cera had scraped off most of the ants by the time Lucy arrived, but their mandibles were still firmly embedded and had to be pulled out one by one. The little girl screamed with each extraction.

Lorcan and Cano had seen these marauders, like a slithering black snake composed of millions of small biting animals, and knew not to get in their way. They watched as a squealing mole rat clawed its way out of a burrow, covered from its eyes to its tail with ants. After making a short dash, the mole rat fell quivering and twitching as more and more ants crawled over it until it was only a bulge in the smooth surface of the column that soon melted away. Predatory ant birds, walking beside the vanguard of the column, snatched up insects and spiders that attempted to escape the ants.

The ants formed their bivouac within the Family's campsite, their nest composed of hundreds of thousands of interlinked bodies. Forced to abandon their camp, the Family moved out of the grove and onto the plain.

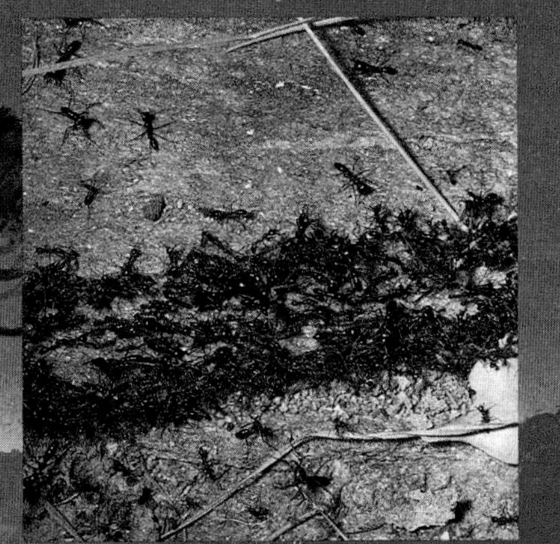
Army ants marching in column

Army-ant bivouac

One of the most feared carnivores in Africa is also the smallest—the army ant.

Hundreds of thousands of these ants stream from underground nests in search of food: animals of any size, living or dead. Using their powerful pincerlike jaws, they rapidly slice up their prey and carry it back to the nest. A column of ants may be a thousand feet long. When army ants are on the move, they construct a new bivouac every night.

Army ants swarming over longhorned grasshopper

Lorcan and the others seemed disheartened as they left their campsite to the rapacious ants. Twilight could be a dangerous time to search for another secure refuge. Eba motioned confidently and emitted a hoarse bark. She strode directly to the far edge of the grove, pointing to a huge rocky outcropping that dominated the flat expanse of the savanna. During her youth her family had spent many secure nights above and between those towering boulders. This kopje would be a far better site than the place they had just abandoned.

Lorcan led the Family swiftly to the summit. Folding down tufts of grass that grew between the boulders, the females made their children comfortable for the night. The older youths and adults stretched out on the warm granite slabs, which still retained the heat of the day. Cano remained at the base of the kopje, watching silently. The dark edge of night crept swiftly across the valley, over the stony fortress, finally extinguishing the last rays of sunset on the still-smoldering volcano.

The prodigal male took his place near the base of the sanctuary as the sounds of day were gradually replaced by those of night. The songs of the birds diminished and blended into a chorus of toads and cicadas. This distant roar of a great cat was met by the lingering howl of a lonesome hyena echoing from the stones. As the moon rose through the brief afterglow, the children's playful chatter gave way to the sounds of sleep.

Eba remained awake while the others slept. All of the Family were near her now. Soon they would all sleep, safe in this rocky eyrie. Eba looked out at the savanna, where the herds moved like great rivers of life. She looked behind her at the smoking mountain that had killed so many that day and had threatened to kill the Family as well. Both the Family and the mountain remained. It was as much a part of their world as were the river and the herds and the moonlight.

The mountain would be there for millions of years. So would the Family.

FURTHER READING

Fossey, Dian. *Gorillas in the Mist.* Boston: Houghton Mifflin Company, 1983.

A scientific and deeply personal view of Fossey's first 13 years studying the mountain gorilla in the Virunga Mountains of Rwanda.

Goodall, Jane. *In the Shadow of Man.* San Diego: San Diego State University Press, 1989.

A detailed but very readable account of Goodall's pioneering fieldwork among the wild chimpanzees at Gombe National Park in Tanzania.

Johanson, Donald C. and Maitland A. Edey. *Lucy: The Beginnings of Humankind.* New York: Warner Books, 1981.

This book is a highly readable account of hominid evolution with detailed discussions of discoveries at Hadar and Laetoli.

Lewin, Roger. *Thread of Life: The Smithsonian Looks at Evolution.* Washington, D.C.: Smithsonian Books, 1982.

A beautifully illustrated account of the diversity of life on earth and the process of evolution.

van Lawick, Hugo. *Among Predators and Prey.* San Francisco: Sierra Club Books, 1986.

Illustrated with stunning photographs, this is a superb example of a true naturalist/photographer's twenty-five years on the Serengeti. The engaging text relates numerous important details about animal behavior.

PHOTO AND

Page xii:
Photo by David Brill

Page 2:
Map by Elizabeth Morales-Denney

Page 3:
Map by Elizabeth Morales-Denney

Page 4:
Photo by David Brill

Page 6:
Photo by Donald C. Johanson

Page 8:
Photos by M. Iwago, National Geographic Society

Page 10:
Illustration by Elizabeth Morales-Denney

Page 12:
Photo by Yann Arthus-Bertrand (Peter Arnold, Inc. MA-AF-4D)

Page 14:
Photos by Donald C. Johanson

Page 15:
(left): Photo courtesy of Robert Harding Picture Library; *(right):* Photo by G. D. Plage (Bruce Coleman, Inc.)

Page 16:
(left): Photo by Yann Arthus-Bertrand (Peter Arnold, Inc. Af 9E-4); *(right):* Photo by Carlo Peretto

Page 23:
Photo by Hugo van Lawick, © National Geographic Society

Page 24:
Photo courtesy of C. Owen Lovejoy

Page 28:
Photo by Mitsuaki Iwago

Page 30:
Photo by Hugo van Lawick, © National Geographic Society

Page 35:
Photos by Hugo van Lawick, © National Geographic Society

Page 37:
(left): Photo by Tim D. White; *(right):* Photo by John Reader. (Photo Researchers, Inc. FA 0251)

Page 38:
(top): Photos © Institute of Human Origins; *(bottom):* Photo by David Brill

Page 39:
Illustration by Elizabeth Morales-Denney

Page 41:
(left): Photo by Philippe Taquet, courtesy of Giancarlo Ligabue; *(center):* Photo by Alexander Marshack; *(right):* Photo by David Brill

Page 43:
(top left): Photo by Yann Arthus-Bertrand (Peter Arnold, Inc.); *(top center):* Photo by Malcolm S. Kirk (Peter Arnold, Inc.); *(top right):* Photo by J. Cancaldsi (Peter Arnold, Inc. Australia 21 B-1); *(bottom left):* Photo by Werner H. Muller (Peter Arnold, Inc. 2512); *(bottom center):* Photo by Carol Beckwith; *(bottom right):* Photo by Malcolm S. Kirk (Peter Arnold, Inc.)

Page 45:
Photo by Donald C. Johanson

Page 47:
(top): Photo by G. Gualco (Bruce Coleman, Inc. 127590); *(bottom):* Photo by Krafft-Explorer (Photo Researchers, Inc. FA 0636)

Page 49:
Photo by Zig Leszczynski (Animals Animals, 415030M)

Page 54:
Photo by Donald C. Johanson

Page 57:
Photo by Hugo van Lawick, © National Geographic Society

Page 58:
Photo by Hugo van Lawick, © National Geographic Society

Page 59:
Photo by Nicholas Devore III (Bruce Coleman, Inc. 606924)

ILLUSTRATION CREDITS

Page 62:
Photo by S. Trevor (Bruce Coleman, Inc. 111234)

Page 65:
(left): Photo by Al Giddings (Ocean Images, Inc.); *(right, top to bottom):* Photo by Donald C. Johanson; Illustration by Elizabeth Morales-Denney; Photo by Donald C. Johanson

Page 67:
(top): Photo by Peter Davey (Bruce Coleman, Inc. 496732); *(bottom):* Photo by Donald C. Johanson

Page 68:
Photo by Clem Haagner (Bruce Coleman, Inc. 625127)

Page 69:
Photo by David Brill

Page 72:
Yann Arthus-Bertrand (Peter Arnold, Inc.)

Page 73:
Illustration by Elizabeth Morales-Denney

Page 77:
Photo by N. Myers (Bruce Coleman, Inc. 9746)

Page 78:
Photos by Donald C. Johanson

Page 80:
Photo by Irv DeVore (Anthro-Photo File)

Page 85:
Photos by David Brill

Page 87:
Photo by Robert M. Campbell, © National Geographic Society

Page 88:
Photo by Anthony Brandenburg (Bruce Coleman, Inc. 462419)

Page 89:
Photo by Mike Price (Bruce Coleman, Inc. 303446)

Page 91:
Photo by Des Bartlett, © *National Geographic* magazine.

Page 92:
(top): Photo by Kim Taylor (Bruce Coleman, Inc. 367203); *(bottom):* Photo by Zig Leszczynski (Animals Animals, R-1850)

Page 93:
(left): Photo by Andrew Odum (Peter Arnold, Inc.); *(center):* Photo by Kim Taylor (Bruce Coleman, Inc. 367162); *(right):* Photo by Donald C. Johanson

Page 95:
Photo by George Schaller (Bruce Coleman, Inc. 460333)

Page 97:
(top): Photo by M. W. Larson (Bruce Coleman, Inc. 59032); *(bottom):* Photo by Edward S. Ross

Page 98:
Photo by Donald C. Johanson

Page 99:
Photo by Lenora Johanson

Page 101:
Illustrations by Elizabeth Morales-Denney

Page 102:
Photo by Hugo van Lawick, © National Geographic Society

Page 105:
Photo by David Brill

Page 107:
Photo by Dieter Blum (Peter Arnold, Inc. NP-US-47A)

Page 109:
Illustration by Elizabeth Morales-Denney

Page 110:
Illustrations by Elizabeth Morales-Denney

Page 111:
Photo by Nicholas Devore III (Bruce Coleman, Inc. 671535)

Page 117
(top left): Photo by Edward S. Ross; *(top right):* Photo by Raymond A. Mendez (Animals Animals, 1-7294); *(bottom):* Photo by Edward S. Ross

ABOUT THE AUTHORS

DR. DONALD JOHANSON, the foremost anthropologist and co-author of the best-seller *Lucy: The Beginnings of Humankind,* is president of the Institute of Human Origins in Berkeley, California. He is donating a portion of the proceeds from this book for the education of Ethiopian students.

KEVIN O'FARRELL, a museum designer, has worked with every major natural history and science museum in North America. He designs many of the offerings of the Nature Company. He lives in Corte Madera, California, and County Kerry, Ireland.